Lecture Notes Electrical Engineering

Volume 49

Subhas Chandra Mukhopadhyay ·
Gourab Sen Gupta · Ray Yueh-Min Huang
(Eds.)

Recent Advances in Sensing Technology

Subhas Chandra Mukhopadhyay
School of Engineering and
Advanced Technology (SEAT)
Massey University (Turitea Campus)
Palmerston North
New Zealand
E-mail: S.C.Mukhopadhyay@
massey.ac.nz

Ray Yueh-Min Huang
Professor and Dept. Chair
Dept. of Engineering Science
National Cheng-Kung University
Tainan, Taiwan
E-mail: huang@mail.ncku.edu.tw

Gourab Sen Gupta
School of Engineering and
Advanced Technology (SEAT)
Massey University (Turitea Campus)
Palmerston North
New Zealand
E-mail: G.Sengupta@massey.ac.nz

ISBN 978-3-642-00577-0   e-ISBN 978-3-642-00578-7

DOI 10.1007/978-3-642-00578-7

Library of Congress Control Number: 2009920581

© 2009 Springer-Verlag Berlin Heidelberg

This work is subject to copyright. All rights are reserved, whether the whole or part of the material is concerned, specifically the rights of translation, reprinting, reuse of illustrations, recitation, broadcasting, reproduction on microfilm or in any other way, and storage in data banks. Duplication of this publication or parts thereof is permitted only under the provisions of the German Copyright Law of September 9, 1965, in its current version, and permission for use must always be obtained from Springer. Violations are liable to prosecution under the German Copyright Law.

The use of general descriptive names, registered names, trademarks, etc. in this publication does not imply, even in the absence of a specific statement, that such names are exempt from the relevant protective laws and regulations and therefore free for general use.

Typeset & Coverdesign: Scientific Publishing Services Pvt. Ltd., Chennai, India.

Printed in acid-free paper

9 8 7 6 5 4 3 2 1

springer.com

Guest Editorial

This Special Issue titled "Recent Advances in Sensing Technology" in the book series of "Lecture Notes in Electrical Engineering" contains the extended version of the papers selected from those that were presented at the 3rd International Conference on Sensing Technology (ICST 2008) which was held in November 30 to December 3, 2008 at National Cheng-Kung University, Tainan, Taiwan. A total of 131 papers were presented at ICST 2008, of which 19 papers have been selected for this special issue.

This Special Issue has focussed on the recent advancements of the different aspects of sensing technology, i.e. information processing, adaptability, recalibration, data fusion, validation, high reliability and integration of novel and high performance sensors. The advancements are in the areas of magnetic, ultrasonic, vision and image sensing, wireless sensors and network, microfluidic, tactile, gyro, flow, surface acoustic wave, humidity, gas, MEMS thermal and ultra-wide band. While future interest in this field is ensured by the constant supply of emerging modalities, techniques and engineering solutions, many of the basic concepts and strategies have already matured and now offer opportunities to build upon.

We do hope that the readers will find this special issue interesting and useful in their research as well as in practical engineering work in the area of sensing technology. We are very happy to be able to offer the readers such a diverse special issue, both in terms of its topical coverage and geographic representation.

Finally, we would like to whole-heartedly thank all the authors for their contribution to this special issue.

Subhas Chandra Mukhopadhyay, Guest Editor
School of Engineering and Advanced Technology (SEAT),
Massey University (Turitea Campus)
Palmerston North, New Zealand
S.C.Mukhopadhyay@massey.ac.nz

Gourab Sen Gupta, Guest Editor
School of Engineering and Advanced Technology (SEAT),
Massey University (Turitea Campus)
Palmerston North, New Zealand
G.SenGupta@massey.ac.nz

Ray Yueh-Min Huang, Guest Editor
Department of Engineering Science
National Cheng-Kung University
Tainan, Taiwan
huang@mail.ncku.edu.tw

VI   Guest Editorial

Dr. Subhas Chandra Mukhopadhyay graduated from the Department of Electrical Engineering, Jadavpur University, Calcutta, India in 1987 with a Gold medal and received the Master of Electrical Engineering degree from Indian Institute of Science, Bangalore, India in 1989. He obtained the PhD (Eng.) degree from Jadavpur University, India in 1994 and Doctor of Engineering degree from Kanazawa University, Japan in 2000.

During 1989-90 he worked almost 2 years in the research and development department of Crompton Greaves Ltd., India. In 1990 he joined as a Lecturer in the Electrical Engineering department, Jadavpur University, India and was promoted to Senior Lecturer of the same department in 1995.

Obtaining Monbusho fellowship he went to Japan in 1995. He worked with Kanazawa University, Japan as researcher and Assistant professor till September 2000.

In September 2000 he joined as Senior Lecturer in the Institute of Information Sciences and Technology, Massey University, New Zealand where he is working currently as an Associate professor. His fields of interest include Sensors and Sensing Technology, Electromagnetics, control, electrical machines and numerical field calculation etc.

He has authored over 200 papers in different international journals and conferences, edited eight conference proceedings. He has also edited four special issues of international journals as guest editor and three books with Springer-Verlag.

He is a Fellow of IET (UK), a senior member of IEEE (USA), an associate editor of IEEE Sensors journal and IEEE Transactions on Instrumentation and Measurements. He is in the editorial board of e-Journal on Non-Destructive Testing, Sensors and Transducers, Transactions on Systems, Signals and Devices (TSSD), Journal on the Patents on Electrical Engineering, Journal of Sensors. He is in the technical programme committee of IEEE Sensors conference, IEEE IMTC conference and IEEE DELTA conference. He was the Technical Programme Chair of ICARA 2004 and ICARA 2006. He was the General chair of ICST 2005, ICST 2007. He has organized the IEEE Sensors conference 2008 at Lecce, Italy as General Co-chair and currently organizing the IEEE Sensors conference 2009 at Christchurch, New Zealand as General Chair.

Dr. Gourab Sen Gupta graduated with a Bachelor of Engineering (Electronics) from the University of Indore, India in 1982. In 1984 he got his Master of Electronics Engineering from the University of Eindhoven, The Netherlands. In 1984 he joined Philips India and worked as an Automation Engineer in the Consumer Electronics division till 1989. Thereafter he worked as a Senior Lecturer in the School of Electrical and Electronic Engineering at Singapore Polytechnic, Singapore. He is with Massey University, Palmerston North, New Zealand, since September 2002 as a Senior Lecturer. He received his PhD in Computer Systems Engineering in 2009 from Massey University, New Zealand. He has published over 80 papers in international journals and conference proceedings, co-authored two books on programming and edited five conference proceedings. He has been a guest editor for leading journals such as IEEE Sensors Journal, International Journal of Intelligent Systems Technologies and Applications (IJISTA), and Studies in Computational Intelligence (Special Issue on Autonomous Robots and Agents) by Springer-Verlag.

His current research interests are in the area of embedded systems, robotics, real-time vision processing, behaviour programming for multi-agent collaboration and sensor applications. He has served on the organising committee of several international conferences.

He is a senior member of IEEE.

Dr. Ray Yueh-Min Huang is a Distinguished Professor and Chairman of the Department of Engineering Science, National Cheng-Kung University, Taiwan, R.O.C. His research interests include Multimedia Communications, Wireless Networks, Embedded Systems, and Artificial Intelligence. He received his MS and Ph.D. degrees in Electrical Engineering from the University of Arizona in 1988 and 1991 respectively. He has co-authored 2 books and has published about 160 refereed professional research papers. He has completed 10 Ph.D. and over 80 MSES thesis students. Dr. Huang has received many research awards, such as the Best Paper Award of 2007 IEA/AIE Conference, Best Paper Award of the Computer Society of the Republic of China in 2003, the Awards of Acer Long-Term Prize in 1996, 1998, and 1999, Excellent Research Awards of National Microcomputer and Communication Contests in 2006. He also received many funded research grants from National Science Council, Ministry of Education, Industrial Technology of Research Institute, and Institute of Information Industry. Dr. Huang has been invited to give talks or served frequently in the program committee at national and international conferences. Dr. Huang is in the editorial board of the Journal of Wireless Communications and Mobile Computing, the Journal of Internet Technology, International Journal of Internet Protocol Technology, International Journal of Ad Hoc and Ubiquitous Computing, Journal of Security and Communication Networks and serves as an associate editor for Journal of Computer Systems, Networks, and Communications as well as International Journal of Communication Systems. He was the Technical Programme Chair of Symposium on Digital Life Technologies (SDLT2007). He was the General chair of VIP2007. He is organizing the SDLT2008, PCM2008 and ICST2008. Huang is a member of the IEEE as well as IEEE communication, computer, and computational intelligence societies.

Contents

Part 1: Magnetic Sensors

A GMR Needle Probe to Estimate Magnetic Fluid Weight
Density Inside Large Tumors 1
*C.P. Gooneratne, A. Kurnicki, M. Iwahara, M. Kakikawa,
S.C. Mukhopadhyay, S. Yamada*

Evaluation of Surface Flaw by Magnetic Flux Leakage
Testing Using Amorphous MI Sensor and Neural Network... 15
M. Abe, S. Biwa, E. Matsumoto

Part 2: Ultrasound/Ultrasonic Sensors

A Scalable Ultrasonic-Based Localization System Using the
Phase Accordance Method..................................... 35
*Toshio Ito, Tetsuya Sato, Kan Tulathimutte, Masanori Sugimoto,
Hiromichi Hashizume*

VHDL-AMS Modelling of Ultrasound Measurement
System in Linear Domain 55
D. Kourtiche, R. Guelaz, A. Rouane, M. Nadi

Application of Laser-Ultrasound to Non-contact
Temperature Sensing of Heated Materials 75
I. Ihara, M. Takahashi, H. Yamada

Part 3: Vision/Image Sensors

Vision Sensor with an Active Digital Fovea 91
Donald G. Bailey, Christos-Savvas Bouganis

Development of a Real-Time Full-Field Range Imaging
System 113
*A.P.P. Jongenelen, A.D. Payne, D.A. Carnegie, A.A. Dorrington,
M.J. Cree*

Development of a Stereo Image Distribution System 131
Junichi Takeno, Toshihiro Enaka, Hirofuji Sato

Part 4: Wireless Sensors and Network

From Labs to Real Environments: The Dark Side
of WSNs.. 143
C. Alippi, C. Camplani, C. Galperti, M. Roveri

Part 5: Microfluidic Sensors

Loading Analysis of a Remotely Interrogatable Passive
Microvalve... 169
Ajay C. Tikka, Said F. Al-Sarawi, Derek Abbott

Part 6: Tactile Sensors

Flexible Piezoelectric Tactile Sensors with Structural
Electrodes Array 189
Cheng-Hsin Chuang

Part 7: Gyro

A New Method for Direct Gravity Estimation and
Compensation in Gyro-Based and Gyro-Free INS
Applications 203
Ehad Akeila, Zoran Salcic, Akshya Swain

Part 8: Flow Sensors

Model-Based Phasor Control of a Coriolis
Mass Flow Meter (CMFM) for the Detection
of Drift in Sensitivity and Zero Point 221
H. Röck, F. Koschmieder

Part 9: Surface Acoustic Wave Sensors

Wireless Interrogation of a Micropump and Analysis of
Corrugated Micro-diaphragms 241
Don W. Dissanayake, Said F. Al-Sarawi, Derek Abbott

Part 10: Humidity Sensors

Synthesis of Aligned ZNO Nanorods with Different Parameters and Their Effects on Humidity Sensing 257
Yun Wang, John T.W. Yeow, Liang-Yih Chen

Part 11: Ultra-Wide Band Sensors

Ultra-Wideband Radars for Through-Wall Imaging in Robotics 271
Jairo Alejandro Gomez, Graham Brooker

Part 12: Gas Sensors

Electrical and Gas Sensing Perfomanance of Coppergermanate 283
V.B. Gaikwad, A.V. Borhade, Y.R. Baste, D.D. Kajale, G.H. Jain

Studies on Gas Sensing Performance of Pure and Nano- Ag Doped ZnO Thick Film Resistors 293
V.B. Gaikwad, M.K. Deore, P.K. Khanna, D.D. Kajale, S.D. Shinde, D.N. Chavan, G.H. Jain

Part 13: MEMS Thermal Sensors

Micro Temperature Sensors and Their Applications to MEMS Thermal Sensors 309
Mitusteru Kimura

Author Index 327

A GMR Needle Probe to Estimate Magnetic Fluid Weight Density Inside Large Tumors

C.P. Gooneratne[1], A. Kurnicki[2], M. Iwahara[1], M. Kakikawa[1],
S.C. Mukhopadhyay[3], and S. Yamada[1]

[1] Kanazawa University, Kanazawa, Japan
[2] Lublin University of Technology, Lublin, Poland
[3] Massey University, Palmerston North, New Zealand

Abstract. In this paper we are proposing a needle-type GMR sensor to estimate magnetic fluid weight density inside large tumors. The application is hyperthermia therapy, a form of cancer treatment. Hyperthermia therapy utilizes the magnetic losses due to magnetization of magnetic nanoparticles by external alternating current magnetic fields. These magnetic losses can be dissipated in the form of heat depending on the thermal conductivity and heat capacity of the surrounding medium. The overall effect is an increase in temperature of the surrounding. This principle is applied to heat and destroy tumors since they are more sensitive to heat than normal healthy cells. Generally all parameters except the magnetic fluid weight density are known in the specific heat equation which governs the heat given in hyperthermia therapy to destroy cancer cells. Hence, accurate estimation of magnetic fluid weight density is critical for successful treatment. This paper presents a methodology to estimate magnetic fluid weight density inside the body and experimental results, by a needle-type GMR sensor.

1 Introduction

Magnetic fluids are colloidal mixtures consisting of superparamagnetic nanoparticles suspended in a carrier fluid, usually an organic solvent or water. Magnetite (Fe3O4) is the most widely used and promising magnetic nanoparticle available today. Since magnetic fluids are stable, colloidal suspensions they possess a unique combination of fluidity and ability to interact and be influenced by magnetic fields. Superparamagnetic nanoparticles have controllable sizes ranging from a few nanometers, which places them at dimensions that are smaller than or comparable to those of a cell (10 – 100 μm), a virus (20 – 450 nm), a protein (5 – 50 nm) or a gene (2 nm wide and 10 – 100 nm long) [1]. Superparamagnetic nanoparticles are non-toxic and biocompatible which means that they are physiologically well tolerated. For example dextran magnetite has no measurable toxicity index LD_{50} [2]. There are many biomedical applications of magnetic fluid due to its special physical properties. Magnetic fluid is used as contrast agents in MRI and coupled with biological molecules and used in cell labeling and separation [3]. Superparamagnetic nanoparticles can also be controlled by an external magnetic field gradient so they are coupled with anti-cancer drugs in targeted drug delivery [4]. Furthermore they resonantly respond to ac magnetic fields, which lead to a transfer of energy from the exciting field to the nanoparticle. This is exploited in hyperthermia therapy where the self heating properties of superparamagnetic nanoparticles can be used to deliver toxic amounts of heat to a tumor [5,6].

Cancer stages are usually categorized according to the size of the tumor and how much it has spread to other tissues and organs [7]. In later stage cancer (stages 3 and 4) the diameter of tumors is 50 mm or more (assuming tumors are spherical shaped). Hyperthermia therapy utilizes heat to destroy tumor cells. Magnetic nanoparticles can be used as self heating agents [8]. Magnetic fluid injected near the affected area inside the body is more readily taken up and hence easily entered into tumor cells compared to healthy cells. Tumor cells are more sensitive to heat than healthy cells [9]. Tumor cells exposed to temperatures in excess of 42.5 °C for a prolonged period of time are destroyed due to apoptosis [9,10]. However, the weight density of the magnetic nanoparticles is an important parameter for giving heat in such a way that it does not affect other healthy cells. Furthermore, magnetic fluid weight density along with applied magnetic flux density amplitude and exciting frequency is directly proportional to the specific heat capacity [11]. Magnetic fluid once injected into the affected area spreads inside tissue effectively reducing its content density, providing an obstacle for successful hyperthermia therapy. This is more complicated in larger cavities since it is difficult to retain magnetic nanoparticles in a large area. By taking into account the difference in magnetic flux density inside a magnetic fluid filled tumor and outside, the fabricated needle-type GMR sensor is proposed to estimate the weight density of the magnetic nanoparticles before and after treatment inside large cavities.

This purpose of this paper is to introduce a uniquely designed novel needle-type GMR sensor for application inside the body in a minimally low-invasive way. A theoretical basis for estimating magnetic fluid weight density inside the body is developed by obtaining a relationship between relative permeability and the magnetic fluid weight density as well as a relationship between the difference of magnetic flux density inside and outside a magnetic fluid filled cavity and the magnetic fluid weight density. An experimental setup, including the needle-type GMR sensor, a Helmholtz tri-coil, and an experimental method in which agar is injected with magnetic fluid to simulate actual clinical process is reported.

2 Analytical Analysis

Magnetic nanoparticles are assumed to be spherical in shape and uniformly distributed in the fluid as shown in figure 1. The relative permeability of magnetic nanoparticles and liquid are assumed to be infinity and one respectively. However, the magnetic nanoparticles have a cluster structure so a spherical structure model is assumed as shown in figure 1. Since there is space between the magnetic nanoparticles the space factor of spherical magnetite is considered and an equation defining the relationship between the relative permeability and weight density is obtained [11], as shown below.

$$\mu^* = 1 + C_d D_w / h_s \gamma_f \qquad (D_w \ll 1) \qquad (2.1)$$

where C_d ; Coefficient (theoretically, 4)
D_w ; Magnetic fluid weight density (mgFe/ml)
h_s ; Space factor of spherical magnetite (0.523)
γ_f ; Specific gravity of magnetite (4.58)

From equation (2.1) it can be seen that the relative permeability is proportional to the magnetic fluid weight density. It is worth to note there is no effect due to the shape or size of the cavity. To confirm equation (2.1) experimental analysis was carried out with the aid of a vibrating sample magnetometer (VSM) as shown in figure 2. When magnetic fluid samples of different weight densities are placed within a uniform magnetic field (provided by the electromagnets in the VSM) and made to undergo sinusoidal motion (mechanically vibrated by the vibration unit) there will be a change in magnetic flux. This induces a voltage in the pick up coils of the VSM, which is proportional to the magnetic moment of the sample. The comparison of the experimental and theoretical results is shown in figure 3. It can be seen that the relative permeability is proportional to the magnetic fluid weight density. The theoretical and experimental results are also in good agreement.

Figure 4 shows a uniform magnetic flux density applied to a spherical cavity filled with magnetic fluid. The magnetic flux lines converge at the magnetic fluid filled cavity. This gives rise to a difference between the applied magnetic flux density B_0, and the magnetic flux density inside B_I, the cavity. The magnetic flux density at the centre of the cavity can be expressed as shown in equation (2.2).

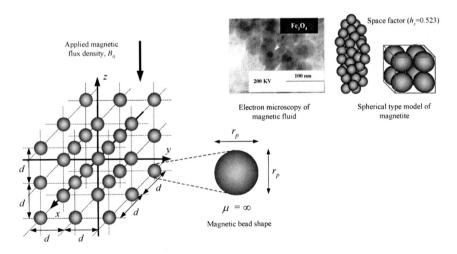

Fig. 1. Magnetic nanoparticle distribution

$$B_1 = \mu^* B_0 / \{1 + N(\mu^* - 1)\} \qquad (\mu^* \approx 1) \qquad (2.2)$$

where N is the demagnetizing factor of the cavity. Equation (2.1) is substituted into (2.2) to obtain the difference δ, between the magnetic flux density inside and outside a magnetic fluid filled cavity.

$$\delta = (B_1 - B_0)/B_0$$
$$\approx \{C_d(1-N)D_w/(h_s\gamma_f)\} \qquad (D_w \ll 1) \qquad (2.3)$$

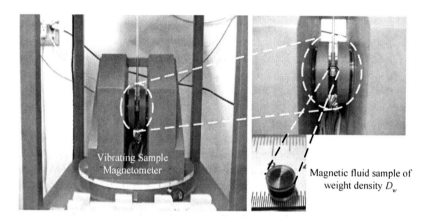

Fig. 2. Experimental setup – Vibrating sample magnetometer

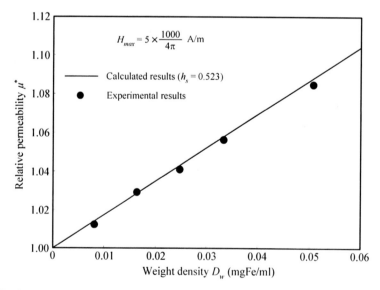

Fig. 3. Relationship between relative permeability and magnetic fluid weight density

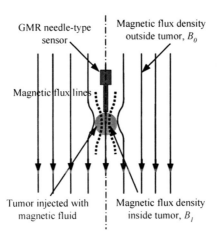

Fig. 4. Spherical magnetic fluid filled cavity under the influence of a uniform magnetic flux density

It can be seen from (2.3) that the magnetic fluid weight density can be calculated from the difference between the magnetic flux density inside and outside a magnetic fluid filled cavity. Magnetic fluid weight density is also proportional to the change ratio of magnetic flux density. The demagnetizing factor N, which depends on the shape and size of the cavity, influences the estimation of magnetic fluid weight density.

3 Needle-Type GMR Sensor

The fabricated needle-type GMR sensor as shown in figure 5 is the key feature of this research. The needle is made of a compound of Aluminium Oxide and Titanium Carbide (Al_2O_3/TiC) and 20 mm in length (15 mm is available to be inserted inside the body). Referring to figure 5 it can be seen that the GMR sensors in the needle probe are designed as a bridge circuit. At the tip of the needle there is a sensing area of 75 × 40 μm. The three other GMR sensors of the bridge circuit are placed near the bonding pads.

Consider the event where the tip of the needle is inserted into the centre of a cavity (as explained in section 2 and figure 4), under a uniform magnetic flux density (B_0). The four GMR sensors are exposed to B_0, assuming that the cavity is empty (permeability of 1 can be assumed inside and outside the cavity). So there is no change in magnetic flux density inside and outside the cavity since B_1 is equal to B_0. However, when the cavity is filled with magnetic fluid the permeability inside is greater than outside the cavity. Hence, the GMR sensing area at the tip of the needle is exposed to a magnetic flux density B_1, which is higher than the applied magnetic flux density B_0. However, since the other three sensors are located further up near the bonding pads, and hence outside the magnetic fluid filled cavity, they will still be exposed to the applied magnetic flux density. This way the magnetic flux density inside a magnetic fluid filled cavity (B_1) and outside the cavity (B_0) can be measured simultaneously.

6   C.P. Gooneratne et al.

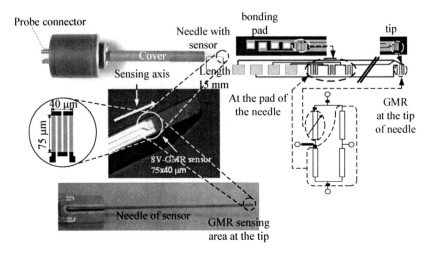

Fig. 5. Needle-type GMR sensor

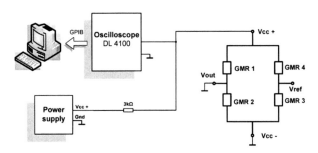

Fig. 6. Experimental setup for small signal ac characterization of needle-type GMR sensor

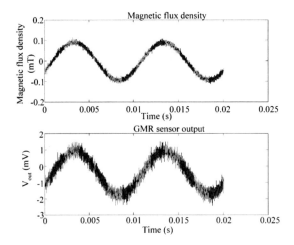

Fig. 7. GMR sensor output for an ac magnetic flux density of 0.1 mT

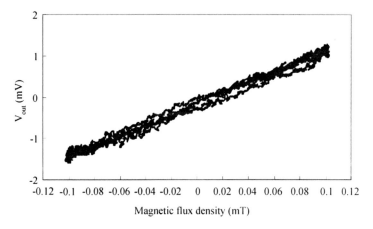

Fig. 8. Small signal ac characterization of GMR sensor at 100 Hz in the sensitive direction

The sensitivity of the needle-type GMR sensor in its sensitive axis was measured as shown in figure 6. The output from the GMR sensing area at the tip as was measured for an ac magnetic flux density of 0.1 mT as shown in figure 7. The results were analyzed by Yokogawa DL4100 oscilloscope and the data was transferred by general purpose interface bus (GPIB) to a personal computer for analysis. The small signal ac characteristics of the needle-type GMR sensor are shown in figure 8. It can be seen that the sensitivity is approximately 13 mV/mT.

4 Measurement Methodology

Since experiments are to be performed with low concentration magnetic fluid it is important that the applied magnetic flux density is at least 1/10th more uniform than the expected change in magnetic flux density inside and outside a magnetic fluid filled cavity. A Helmholtz tri-coil [12,13] as shown in figure 9 is fabricated to provide a uniform magnetic flux density. The Helmholtz tri-coil produces a uniform magnetic flux density of 0.1 mT at 100 Hz for experiments. The contour plot of error less than 0.01 % is shown in figure 10. The fluctuation of the magnetic flux density is 0.01 %, 0.03 m in the axial and radial direction from the midpoint.

The experiment method for small cavities simulating 1/2nd stage cancerous tumors (less than 20 mm diameter) involves inserting the sensor needle at the centre of the cavity, where the magnetic flux density is most uniform. The experimental method for large cavities [13] is shown in figure 11. Since the sensor needle is only 15 mm in length it cannot be inserted to the centre of large cavities (larger than 30 mm in height). Also stage 3-4 cancerous tumors are generally more than 50 mm in diameter. The method proposed involves estimating the change in magnetic flux density at 20 mm steps (approximate distance between the sensing area at the tip and the three sensing areas near the bonding pads of the needle-type probe) as shown in figure 11. The total change in the magnetic flux density is calculated by summing the change in magnetic flux densities at each step as shown below [13].

$$B_{TotalChange} = \sum_{i=0}^{n}(B_{i+1} - B_i) \qquad (4.1)$$

Fig. 9. Helmholtz tri-coil

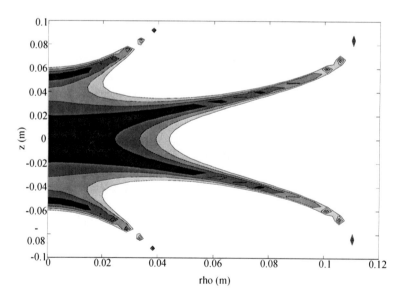

Fig. 10. Contour error plot (0.01 %) for Helmholtz tri-coil

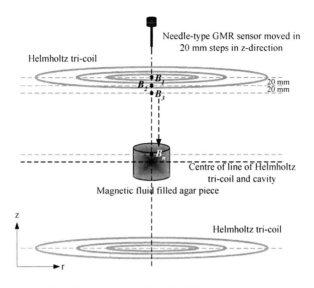

Fig. 11. Experimental method for large cavities

5 Experimental Results

The experimental setup is shown in figure 12. The Sony Tektronix AF5310 function generator provided the ac current at 100 Hz frequency. The function generator was connected to a NF Electronic Instrument 4055 high speed amplifier and the output was fed into the Helmholtz tri-coil. A current clamp probe connected to a Hioki 3272 power supply was connected to the Helmholtz tri-coil so that the current waveform could be analyzed using the Yokogawa DL4100 oscilloscope. The results from the bridge output of the sensor were analyzed through the NF electronics LI5640 lock-in amplifier.

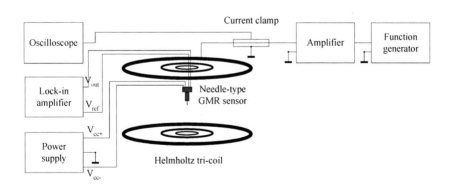

Fig. 12. Experimental setup

The demagnetizing factor N, of a cavity depends on the aspect ratio s. The aspect ratio of a cavity s = (long axis/diameter). Cylindrical agar cavities of diameter 18 and 63 mm (s = 1) were injected with magnetic fluid. The needle-type sensor is then applied as explained in section 2 for 18 mm cavities and section 4 for 63 mm cavities. The experimental results are shown in figure 13. It can be seen that the total change in the magnetic flux density increases with weight density of the magnetic fluid. Moreover, the experimental results for large cavities compare favourably with theoretical results based on ellipsoidal cavities, numerical results obtained by numerical modelling and experimental results obtained for small cavities.

Figure 14 shows the results obtained for different size cylindrical agar cavities (s = 1) for a weight density of 2.286 %. In previous experiments done on smaller cavities [14], equation (2.3) has been verified for a range of sizes. It was shown that the change in magnetic flux density did not vary so much between different size cavities as long as s and hence N remained the same. The change in magnetic flux density only increased with increasing magnetic fluid weight density. However, it must be noted that these experiments were performed with cavities smaller than 30 mm in diameter, hence allowing the insertion of the needle tip to the centre of the cavity. In the experiments performed in this paper the larger the cavity the further away it would be from the centre of the cavity (where the magnetic flux density is the greatest) as shown in table 1. Thus, as diameter and/or

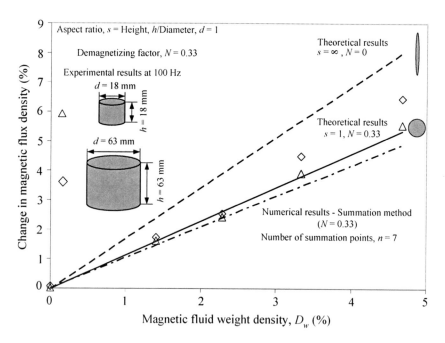

Fig. 13. Estimation of magnetic fluid weight density inside large cavities by measuring the change in magnetic flux density by the needle-type GMR sensor

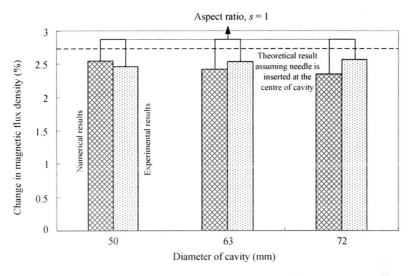

Fig. 14. Estimation of magnetic fluid weight density in different sized cavities for $D_w = 2.29\ \%$

Table 1. Comparison of different sized cavities when needle is completely inserted inside the cavity

Cavity	Diameter (mm)	Height (mm)	Distance from centre of cavity to needle tip (mm)
1	50	50	10.0
2	63	63	16.5
3	72	72	21.0

height of a cavity increases (for a constant N) the total change in magnetic flux density decreases. Figure 14 compares the experimental results to numerical results. Even though the experimental results increase with the size of cavity they do not fluctuate so much compared to the numerical results.

6 Conclusion

The application of magnetic principles in medicine has paved the way for safe, low-cost, non-invasive/minimally invasive, healthcare for patients worldwide. Superparamagnetic nanoparticles or magnetic fluid is one of the most promising advancements in the application of magnetism in biomedicine. The main feature of magnetic fluid is that it can be controlled by an external magnetic field which opens up limitless opportunities in applications such as biomolecular tagging, targeted drug delivery, MRI and hyperthermia. Magnetic fluid hyperthermia has the potential to be an effective, non-invasive cancer therapy with negligible side effects. Magnetic fluid is injected into the affected area and an external ac magnetic field is added to exploit the self heating properties of the particles.

One of the main problems associated with magnetic fluid hyperthermia is that the magnetic fluid spreads inside tissue once injected, reducing its content density. Specific heat capacity is directly proportional to magnetic fluid content density. Inaccurate estimation of magnetic fluid content density has two major effects; i) if a low dosage is given the overall effect is thermal under-dosage in the target region which often leads to recurrent tumor growth ii) if heat given to a target region exceeds the therapeutic limit it may damage healthy cells. Hence, it can be stated that the quality of magnetic fluid hyperthermia treatment is proportional to the accuracy of estimating magnetic fluid content density inside the body. Hyperthermia therapy is most effectively performed on primary tumors. The diameter of primary tumors is typically less than 20 mm, assuming a spherical shape. Tumor sizes in stages 3 and 4 of cancer are more than 50 mm in diameter further complicating matters in retention of magnetic fluid. A novel needle-type GMR sensor was fabricated and utilized to estimate the magnetic fluid weight densities in large cavities. Theoretical analysis presenting the basis for estimating magnetic fluid weight density was presented.

Experiments were performed with agar cavities simulating cancerous tumors at various stages. The different size agar cavities were injected with magnetic fluid to simulate fluid filled tumors in cancer. The needle-type GMR sensor was then used to measure the change in the magnetic flux density. Since the cylindrical agar cavities representing tumors in the $1/2^{nd}$ stage cancer were 18 mm the 20 mm needle was easily inserted into the centre to measure magnetic flux density inside the cavity. However, cavities which represent $3/4^{th}$ stage tumors were 63 mm so the needle could not be inserted into the centre. Hence, a new summing method was developed to estimate magnetic fluid weight density by differential magnetic flux density. The sensing area at the tip and the bridge circuit of the design was then exploited and the change in magnetic flux density was estimated at a range of points and then summed to get the total change for a range of magnetic fluid weight densities. The results show that the change in magnetic flux density is proportional to the magnetic fluid weight density and the results fall between the theoretical lines for $N = 0$ and 0.33 based on ellipsoidal cavities. Furthermore, it has good agreement with experimental results for small cavities of same demagnetizing factor and numerical results that simulated the actual experimental condition. Further experiments were done to measure the magnetic fluid weight density in different sized cavities for constant N. Currently hyperthermia therapy for large tumors is performed mainly to kill part of the tumor since it is difficult to accurately measure the magnetic fluid weight density inside the tumor. This makes it difficult to provide accurate heat to kill tumors and not affect healthy cells. Usually further chemotherapy and/or radiotherapy is performed in lower doses after hyperthermia therapy to completely kill the tumor. By precise estimation of magnetic fluid weight density by inside large tumors hyperthermia therapy has a good potential to be an effective treatment.

The fabricated needle-type GMR sensor has a good possibility to be used for such purpose in the future. Some positive aspects of using the method proposed can be taken into account when considering actual clinical procedure. The needle sensor can be inserted into the body in a minimally low-invasive way. The needle-type GMR sensor has a good potential to be used in other medical applications such as targeted drug delivery where the needle-sensor can be used to confirm if the magnetic particles and drugs are present at a given site by detection as well as estimation of the content density for supplying heat to initiate a chemical reaction. Due considerations should also be given to improve some aspects of the proposed method for successful implementation in clinical applications. While the needle currently used is thin enough to be minimally invasive, it is fragile and is thus uneasy for inserting inside the human body. A needle must be designed and fabricated that would be safe to be used inside the human body. A design in the form of an acupuncture needle has the possibility of being minimally invasive as well as safe to be used inside the body. Currently it is assumed that tumors are spherical in shape. However, in reality they could be of any shape. This gives rise to the situation where the orientation of the cavity should be considered. Hence, more experiments are needed to be performed with different shapes and orientations of cavities with regards to the actual situation.

Acknowledgement

This research was supported in part by the Foundation for Japan Science and Technology Agency, 07-029.

References

[1] Pankhurst, Q.A., Connoly, J., Jones, S.K., Dobson, J.: Applications of magnetic nanoparticles in biomedicine. Journal of Physics D: Applied Physics 36, R167–R181 (2003)
[2] Babincova, M., Leszczynska, D., Sourivong, P., Babinec, P.: Blood-specific whole-body electromagnetic hyperthermia. Medical Hypotheses 55(6), 459–460 (2000)
[3] Berry, C.C., Curtis, A.S.G.: Functionalisation of magnetic nanoparticles for applications in biomedicine. Journal of Physics D: Applied Physics 36, R198–R206 (2003)
[4] Gupta, A.K., Gupta, M.: Synthesis and surface engineering of iron oxide nano particles for biomedical applications. Biomaterials 26, 3995–4021 (2005)
[5] Jordan, A., Scholz, R., Maier-Hauff, K., Johannsen, M., Wust, P., Nadobny, J., Schirra, H., Schmidt, H., Deger, S., Loening, S., Lanksch, W., Felix, R.: Presentation of a new magnetic field therapy system for the treatment of human solid tumors with magnetic fluid hyperthermia. Journal of Magnetism and Magnetic Materials 225, 118–126 (2001)
[6] Jordan, A., Scholz, R., Wust, P., Fähling, H., Felix, R.: Magnetic fluid hyperthermia (MFH): Cancer treatment with AC magnetic field induced excitation of biocompatible superparamagnetic nanoparticles. Journal of Magnetism and Magnetic Materials 201, 413–419 (1999)
[7] Hilger, I., Hergt, R., Kaiser, W.A.: Towards breast cancer treatment by magnetic heating. Journal of Magnetism and Magnetic Materials, 314–319 (2005)

8. Bae, S., Lee, S.W.: Applications of NiFe$_2$O$_4$ nanoparticles for a hyperthermia agent in biomedicine. Applied Physics Letters 89, 252503 (2006)
9. Mukhopadhyay, S.C., Chomsuwan, K., Gooneratne, C.P., Yamada, S.: A Novel Needle-Type GMR Sensor for Biomedical Applications. IEEE Sensors Journal 31, 401–408 (2007)
10. Yamada, S., Chomsuwan, K., Mukhopadhyay, S.C., Iwahara, M., Kakikawa, M., Nagano, I.: Detection of Magnetic Fluid Volume Density with a GMR Sensor. Journal of the Magnetics Society of Japan 31, 44–47 (2007)
11. Gooneratne, C., Chomsuwan, K., Łękawa, A., Kakikawa, M., Iwahara, M., Yamada, S.: Estimation of Density of Low-Concentration Magnetic Fluid by a Needle-Type GMR Sensor for Medical Applications. Journal of the Magnetics Society of Japan 31, 191–194 (2008)
12. Sasada, I., Nakashima, Y.: A planar coil system consisting of three coil pairs for producing uniform magnetic field. Journal of Applied Physics 99(8), 08D904–08D904-3 (2006)
13. Gooneratne, C.P., Iwahara, M., Kakikawa, M., Yamada, S., Kurnicki, A., Mukhopadhyay, S.C.: Magnetic fluid weight density estimation in large cavities by a needle-type GMR sensor. In: Proceedings of the 3rd International Conference on Sensing Technology 2008, pp. 642–647. IEEE, Los Alamitos (2008)
14. Gooneratne, C., Łękawa, A., Iwahara, M., Kakikawa, M., Yamada, S.: Estimation of Low-Concentration Magnetic Fluid Weight Density and Detection inside an Artificial Medium Using a Novel GMR Sensor. Sensors and Transducers Journal 90 (Special issue), 27–38 (2008)

Evaluation of Surface Flaw by Magnetic Flux Leakage Testing Using Amorphous MI Sensor and Neural Network

M. Abe, S. Biwa, and E. Matsumoto

Kyoto University,
Kyoto, Japan

Abstract. In this paper, we attempt to evaluate the shape of a surface flaw including its horizontal position and the located surface by biaxial Magnetic Flux Leakage Testing (MFLT) using an amorphous MI sensor and a Neural Network (NN). The specimen is a magnetic material subjected to the magnetic field, and the Magnetic Flux Leakage (MFL) occurs near the flaw. We measure the biaxial MFL, i.e., the tangential and the normal components of MFL by an amorphous MI sensor. The amorphous MI sensor has the wide measurement range, high sensitivity and high spacial resolution, so that it is suitable for precise quantitative evaluation of the flaw by MFLT. Initially, we pre-process the measured biaxial MFL by Regression Analysis Method (RAM) to extract MFL parameters. Subsequently, NN is used to infer the dimension of the cross section of the flaw including its horizontal position and the located surface from the MFL parameters. By repeating a similar process along several measurement lines parallel to the specimen surface, we can identify the three-dimensional shape of the flaw. In this paper, we first evaluate three-dimensional shape of a parallelepiped flaw in SS400 specimen as the simplest case. Secondly, we consider extending the evaluation method to an oblique flaw based on the two-dimensional magnetostatic analysis by use of Finite Element Method (FEM). The results show that the three-dimensional shape of a parallelepiped flaw and the two-dimensional shape of an oblique flaw can be evaluated with good accuracy.

1 Introduction

Magnetic Flux Leakage Testing (MFLT) is a commonly used Non-Destructive Testing (NDT) technique for inspection of gas or oil pipe lines, and it is also promising for structures such as rail tracks, oil storage tank floors etc. [1]. If there exists a flaw on the surface of a ferromagnetic material under the magnetic field, the magnetic flux leaks from the material due to the variation of the magnetic property around the flaw. MFLT evaluates the flaw by means of the correlation between the flaw shape and the distribution of the Magnetic Flux Leakage (MFL) near the flaw measured by a magnetic sensor. Recent development in electronics and sensor technique enable magnetic sensors to be smaller and more sensitive. In this paper we use newly developed amorphous MI sensor, which has very small sensing element and high resolution [2,3]. Such kind of small and high-sensitive magnetic sensors are applied in MFLT and make it prospective. However, a definitive method for quantitative evaluation of flaw shape by MFLT is not established yet. This is because the most important problem in MFLT is the signal analysis of the measured MFL [4,5]. In other words, the correlation between the flaw shape and the MFL is usually so complex that utilizing it is not easy. In this paper, we use Neural Network (NN) to obtain the correlation.

For training of NN, the MFL distribution near the flaw is employed as the input and the flaw shape as the output. In many cases, three-dimensional flaw shapes have been evaluated from planar MFL distributions [6,7]. However, this kind of method needs a great amount of training data sets for a large variety of flaw shapes. So we adopt a method where a cross section of a flaw can be evaluated from one-dimensional MFL distribution on a measurement line. The three-dimensional flaw shape is evaluated by repeating the above procedure along several measurement lines. This method can significantly reduce the training data sets for NN.

On the other hand, for improvement of leaning efficiency and accuracy of reconstruction, it is feasible to extract a certain number of Characteristic Quantities (CQs) from the MFL. We propose the Regression Analysis Method (RAM) as a new method for characterizing the MFL distribution [8]. We approximate the original MFL distribution by elementary functions with a small number of CQs. Since the analytical MFL distribution for a two-dimensional flaw can be expressed in terms of elementary functions based on the magnetic dipole model [9,10], the proposed RAM may characterize the measured MFL distribution with good accuracy.

In this paper, we first introduce the elementary functions for a rectangular flaw and an oblique flaw based on the magnetic dipole model. Subsequently we evaluate three-dimensional shape of a parallelepiped flaw in SS400 specimen by adopting the RAM as the pre-processing method and the NN as the inverse analysis method. Finally, we consider extending the evaluation method to an oblique flaw based on the two-dimensional magnetostatic analysis with Finite Element Method (FEM).

2 Distribution Functions of MFL

In the RAM proposed in the previous section, CQs are extracted from the MFL distribution by use of elementary distribution functions. In this section, the elementary functions are derived from the analytical expressions based on the magnetic dipole model for simple shape flaws.

First, we consider two-dimensional flaw with a rectangular cross section as the simplest case, refer to figure 1. The dipole model assumes that positive and negative charges are uniformly located on each side surface of the flaw, where the flaw locates at the origin of the coordinate axes and extends infinitely along y axis. Then the tangential and normal components B_x and B_z of the MFL without the external magnetic field can be expressed in the following form [9,10].

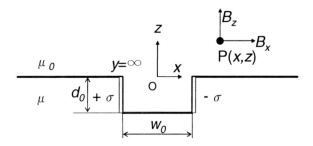

Fig. 1. Rectangular flaw parameters

$$B_x = 2\sigma\mu_0 \left[\tan^{-1} \frac{d_0(x+w_0/2)}{(x+w_0/2)^2 + z(z+d_0)} - \tan^{-1} \frac{d_0(x-w_0/2)}{(x-w_0/2)^2 + z(z+d_0)} \right] \quad (1)$$

$$B_z = \sigma\mu_0 \ln \left[\frac{(x+w_0/2)^2 + (z+d_0)^2}{(x-w_0/2)^2 + (z+d_0)^2} \frac{(x-w_0/2)^2 + z^2}{(x+w_0/2)^2 + z^2} \right] \quad (2)$$

where d_0 is the depth, w_0 the width and σ(Wb/m^2) is the surface magnetic density. If the flaw width w_0 is sufficiently small, the above expressions are approximated as

$$B_x = 2\sigma\mu_0 w_0 \left[\frac{z}{x^2+z^2} - \frac{z+d_0}{x^2+(z+d_0)^2} \right] \quad (3)$$

$$B_z = -2\sigma\mu_0 w_0 \left[\frac{x}{x^2+z^2} - \frac{x}{x^2+(z+d_0)^2} \right] \quad (4)$$

The above expressions are identical to the Forster's MFL distribution functions for a finite slot [11]. Since in a realistic case the horizontal position of the flaw is unknown and there exists the external magnetic field, we should modify the above expressions. Furthermore we are interested in the MFL distribution at a height, i.e., the lift-off of the magnetic sensor. In a result, the one-dimensional MFL distributions at the constant lift-off are expressed in the form.

$$B_x = a_x \left[\frac{b_x}{(x-d_x)^2 + b_x^2} - \frac{c_x}{(x-d_x)^2 + c_x^2} \right] + e_x \quad (5)$$

$$B_z = -a_z \left[\frac{x-d_z}{(x-d_z)^2 + b_z^2} - \frac{x-d_z}{(x-d_z)^2 + c_z^2} \right] - e_z(x-d_z) + f_z \quad (6)$$

Subsequently, as the next step, we assume that the flaw is inclined at θ to the normal direction of the specimen surface and extends infinitely along y axis, refer to figure 2. According to the dipole model, the two-dimensional MFL distributions B_x and B_z without the external magnetic field can be expressed in the following form [9,10].

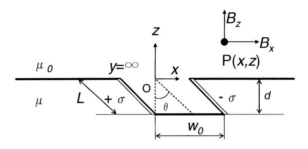

Fig. 2. Oblique flaw parameters

$$B_x = 2\sigma\mu_0 \left[\cos\theta \tan^{-1} \frac{L\{(x+w_0/2)\cos\theta + z\sin\theta\}}{(x+w_0/2)^2 + z^2 + L\{z\cos\theta - (x+w_0/2)\sin\theta\}} \right.$$
$$\left. - \frac{\sin\theta}{2} \ln \frac{(x+w_0/2)^2 + z^2 + L^2 + 2L\{z\cos\theta - (x+w_0/2)\sin\theta\}}{(x+w_0/2)^2 + z^2} \right]$$
$$- 2\sigma\mu_0 \left[\cos\theta \tan^{-1} \frac{L\{(x-w_0/2)\cos\theta + z\sin\theta\}}{(x-w_0/2)^2 + z^2 + L\{z\cos\theta - (x-w_0/2)\sin\theta\}} \right.$$
$$\left. - \frac{\sin\theta}{2} \ln \frac{(x-w_0/2)^2 + z^2 + L^2 + 2L\{z\cos\theta - (x-w_0/2)\sin\theta\}}{(x-w_0/2)^2 + z^2} \right] \quad (7)$$

$$B_z = \mu_0\sigma \left[\cos\theta \ln \frac{(x+w_0/2)^2 + z^2 + L^2 + 2L\{z\cos\theta - (x+w_0/2)\sin\theta\}}{(x+w_0/2)^2 + z^2} \right.$$
$$\left. + 2\sin\theta \tan^{-1} \frac{L\{(x+w_0/2)\cos\theta + z\sin\theta\}}{(x+w_0/2)^2 + z^2 + L\{z\cos\theta - (x+w_0/2)\sin\theta\}} \right]$$
$$- \mu_0\sigma \left[\cos\theta \ln \frac{(x-w_0/2)^2 + z^2 + L^2 + 2L\{z\cos\theta - (x-w_0/2)\sin\theta\}}{(x-w_0/2)^2 + z^2} \right.$$
$$\left. + 2\sin\theta \tan^{-1} \frac{L\{(x-w_0/2)\cos\theta + z\sin\theta\}}{(x-w_0/2)^2 + z^2 + L\{z\cos\theta - (x-w_0/2)\sin\theta\}} \right] \quad (8)$$

where w_0 is the width and L the depth in the direction of θ. If the flaw width w_0 is sufficiently small, the above expressions can be approximated as

$$B_x = 2\sigma\mu_0 w_0 \left[\cos\theta \left\{ \frac{z}{x^2+z^2} - \frac{z+L\cos\theta}{(x-L\sin\theta)^2 + (z+L\cos\theta)^2} \right\} \right.$$
$$\left. - \sin\theta \left\{ \frac{x-L\sin\theta}{(x-L\sin\theta)^2 + (z+L\cos\theta)^2} - \frac{x}{x^2+z^2} \right\} \right] \quad (9)$$
$$= 2\sigma\mu_0 w_0 \cos\theta \left[\frac{z+x\tan\theta}{x^2+z^2} - \frac{z+L\cos\theta + (x-L\sin\theta)\tan\theta}{(x-L\sin\theta)^2 + (z+L\cos\theta)^2} \right]$$

$$B_z = \mu_0 \sigma w_0 \left[2\cos\theta \left\{ \frac{x - L\sin\theta}{(x - L\sin\theta)^2 + (z + L\cos\theta)^2} - \frac{x}{x^2 + z^2} \right\} \right.$$

$$\left. + 2\sin\theta \left[\frac{z}{x^2 + z^2} - \frac{z + L\cos\theta}{(x - L\sin\theta)^2 + (z + L\cos\theta)^2} \right] \right] \quad (10)$$

$$= -2\mu_0 \sigma w_0 \cos\theta \left[\frac{x - z\tan\theta}{x^2 + z^2} - \frac{(x - L\sin\theta) - (z + L\cos\theta)\tan\theta}{(x - L\sin\theta)^2 + (z + L\cos\theta)^2} \right]$$

where $L\cos\theta$ is the vertical depth of the flaw. Similarly to the forgoing case, we can derive from the above equations the one-dimensional MFL distributions at the constant lift-off expressed in the form.

$$B_x = a_x \left[\frac{b_x + h_x(x - d_x)}{(x - d_x)^2 + b_x^2} - \frac{c_x + h_x(x - d_x - i_x)}{(x - d_x - i_x)^2 + c_x^2} \right] + e_x \quad (11)$$

$$B_z = -a_z \left[\frac{(x - d_z) - h_z b_z}{(x - d_z)^2 + b_z^2} - \frac{(x - d_z - i_z) - h_z c_z}{(x - d_z - i_z)^2 + c_z^2} \right] - e_z(x - d_z) + f_z \quad (12)$$

where h and i are the parameters depending on the flaw angle θ. If $\theta = 0$, the above formulas become equations (5) and (6).

3 Characteristic Quantities Extracted by Ram

As a procedure of RAM, we approximate the MFL distributions by the elementary functions introduced in the previous section with the least square method, where CQs are the coefficients contained in the functions. In figure 3, circles denote the tangential and the normal components of the measured MFL distribution for a rectangular flaw. The solid lines in figure 3 show the approximated MFL distributions of equations (5) and (6). Likewise, Figure 4 shows the calculated MFL distribution for an oblique flaw and the result of approximation by use of equations (11) and (12). In each figure, the lines are properly coincident with the measured data. In equations (5), (6), (11) and (12) parameters d_x and d_z indicate the horizontal position of the flaw. Parameters e_x, e_z and f_z are the tangential and normal components of the bias magnetic flux far from the flaw. The above parameters can be determined in the approximation process of the measured MFL distribution. Thus, we employ the remained parameters as the CQs to evaluate the flaw shape quantitatively in the following section. In the case of a rectangular flaw, CQs are the six parameters a_x, b_x, c_x, a_z, b_z, c_z. Similarly, in the case of an oblique flaw, CQs are the ten parameters a_x, b_x, c_x, h_x, i_x, a_z, b_z, c_z, h_z, i_z.

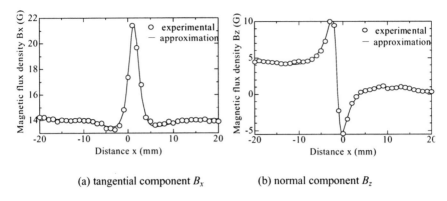

Fig. 3. Approximation of biaxial MFL distributions of rectangular flaw

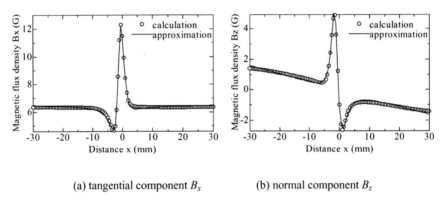

Fig. 4. Approximation of biaxial MFL distributions of oblique flaw

4 MFL Measurement of Parallelepiped Flaw

For the measurement of the biaxial MFL distributions, i.e., B_x and B_z, we use an amorphous MI sensor (MI-CB-1DW, AMI Co., figure 5) with high spacial resolution achieved by 0.45 mm square sensing element. The sensor also has large measurement range ±30 G and high resolution 1 mG. The specimen material is conventional structural rolled steel SS400. Each specimen is 300 mm length, 30 mm width and 5 mm thickness, and it has parallelepiped flaws at sufficiently intervals on one surface, refer to figure 6. For training set of the NN, we introduced 24 flaws with width w and depth d indicated by solid circles in figure 7. We measure the biaxial MFL distributions on the front and the back surfaces with respect to a magnetic sensor, so that we obtain 48 training data sets. Meanwhile, for evaluation of unknown flaws, we prepare 6 front and back surface flaws with width w and depth d given by blank circles in figure 7.

Evaluation of Surface Flaw by MFLT Using Amorphous MI Sensor and NN 21

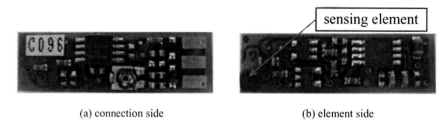

(a) connection side   (b) element side

Fig. 5. Photo of amorphous MI sensor

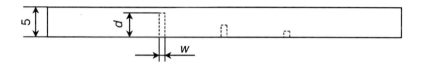

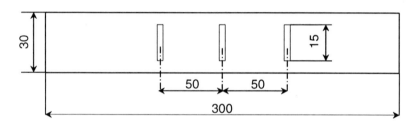

Fig. 6. Specimen (SS400 steel)

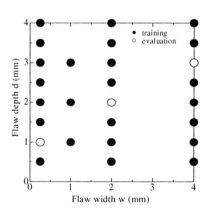

Fig. 7. Widths and depths of prepared parallelepiped flaws

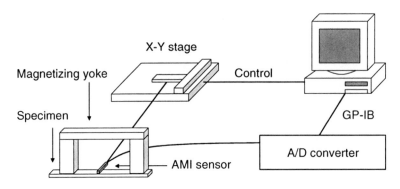

Fig. 8. Experimental system

Figure 8 shows the schematic diagram of the experimental system. The specimen is magnetized by a magnetizing yoke (HMA-1, Eishin Kagaku Co.) whose magnetic core has 25 mm square cross section and 140 mm distance poles. DC current of 0.3 A is supplied to the magnetizer such that 0.6 T magnetic flux density is induced in the cross section of the specimen far from the flaw. The magnetic flux density is measured by an amorphous MI sensor. Throughout the experiment, we fix the lift-off of the sensor at 1.0 mm and measure the biaxial MFL with 1 mm pitch along the 40 mm line including the flaw. We next repeat the same procedure along the lines with 1 mm intervals in the 20 mm range including the flaw. Thus we measured the biaxial MFL over the 40 mm × 20 mm area at lattice points with 1 mm intervals for each one of 54 flaws.

5 Training of NN for Parallelepiped Flaw

For the training of the NN to evaluate shapes of a parallelepiped flaw, we use the 48 data sets measured in the previous section. Each data set consists of 20 one-dimensional biaxial MFL distributions, i.e., we obtained 960 one-dimensional biaxial MFL data. These 960 data include the data for lines under which the flaw is present and for lines under which the flaw is not present. According to discussions in sections 2 and 3, we obtain all the parameters appearing in equations (5) and (6) for the known flaws.

First, we construct the NN to discriminate whether the flaw is present under the each measurement line or not. This NN classify data for each measurement line into defective or non-defective. We call this NN the classification-NN. In the classification-NN, the inputs are the six CQs a_x, b_x, c_x, a_z, b_z, c_z and the outputs are symbols d and nd which represent whether a flaw exists, refer to figure 9. Here and henceforth symbols d and nd denote the defect and non-defect, respectively. In this paper, we use NN software, Neural Works Predict, Neural Ware.

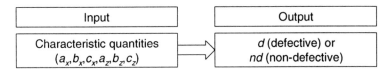

Fig. 9. Structure of classification-NN for evaluation of parallelepiped flaw

Subsequently, we construct the NN to judge the located surface of the flaw. We call this NN the localization-NN. For training of this NN, we use the data only for lines under which the flaw is present. In the localization-NN, the inputs are the above six CQs and the outputs are the symbols f and b which represent the located surface, refer to figure 10. Here and henceforth symbols f and b denote the front and back surface, respectively.

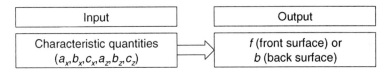

Fig. 10. Structure of localization-NN for evaluation of parallelepiped flaw

Finally, we construct the NN to identify the cross section of a flaw, i.e., the width and the depth. We call this NN the identification-NN. For training of this NN, we use the data only for lines under which the flaw is present, similarly to the localization-NN. In the identification-NN, the inputs are the above six CQs and the outputs are the width and the depth of the flaw, refer to figure 11.

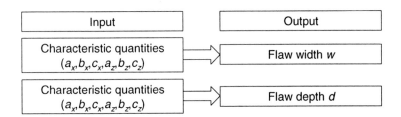

Fig. 11. Structure of identification-NN for evaluation of parallelepiped flaw

6 Evaluation of Unknown Parallelepiped Flaw

For the evaluation of unknown parallelepiped flaws by the above trained NN, we use the data for 6 front and back surface flaws measured in section 4.

We evaluate the flaws in three steps. In the first step, we estimate whether the data for each measurement line is defective or not by the classification-NN. In the second step, we evaluate the located surface with respect to the "defective" data

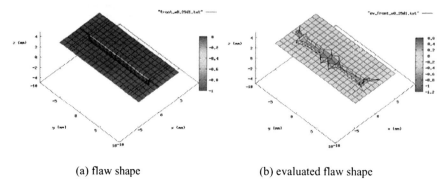

(a) flaw shape (b) evaluated flaw shape

Fig. 12. Evaluation of front surface flaw with 0.25mm width and 1mm depth

Table 1. Errors in evaluation of front surface flaw with 0.25mm width and 1mm depth

	Average error	Maximum error
Defective or non-defective	85.7% accuracy	
Located surface	100% accuracy	
Width	0.98mm	2.48mm
Depth	0.35mm	0.44mm

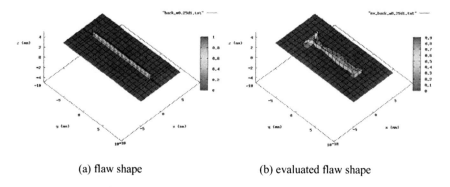

(a) flaw shape (b) evaluated flaw shape

Fig. 13. Evaluation of back surface flaw with 0.25mm width and 1mm depth

Table 2. Errors in evaluation of back surface flaw with 0.25mm width and 1mm depth

	Average error	Maximum error
Defective or non-defective	90.5% accuracy	
Located surface	100% accuracy	
Width	0.89mm	3.67mm
Depth	0.26mm	0.38mm

which is classified as defective in the first step. In this step, we use the localization-NN. In the final step, we identify the width and the depth of the cross section of the flaw, by use of the identification-NN. In this step, we evaluate only the "defective" data, similarly to the second step. Besides, we adopt the average value of the d_x and d_z, which are obtained at the point of extraction of CQs by RAM, as the horizontal position of the flaw.

In figures 12 to 17, figures (a) show the flaw shapes and figures (b) show the evaluated flaw shapes. In the figures, the graduated value denotes the depth (mm), where the negative and positive values represent the front surface and the back surface flaws, respectively. Table 1 to 6 show the averaged error and the maximum error on the 15 measurement lines cover the opening area of each flaw.

The accuracy of distinguishing between the defective lines and the non-defective lines by the classification-NN is between 80 ~ 95 %. The accuracy of estimation of the located surface by the localization-NN is over 93 %. These results show it is not likely that the flaw is overlooked or the located surface is wrongly estimated if the flaw passes over several measurement lines. On the other hand, the averaged errors in evaluations of the width and the depth are around 0.1 ~ 1 mm. The range of above errors is in a similar order to the sensing element size or the scanning intervals. And it can be seen that the maximum error in evaluation of the width tends to be larger as the width become smaller. The averaged error and the maximum error in evaluations of the depth are around 0.3 mm and 0.6 mm, respectively. Since the most important parameter for the materials diagnostics is the flaw depth, it can be said the above evaluation accuracy is enough good for practical use in material inspection.

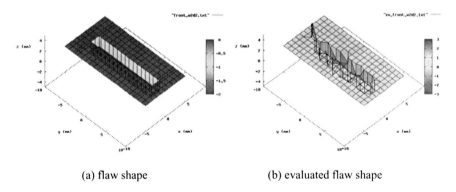

(a) flaw shape   (b) evaluated flaw shape

Fig. 14. Evaluation of front surface flaw with 2mm width and 2mm depth

Table 3. Errors in evaluation of front surface flaw with 2mm width and 2mm depth

	Average error	Maximum error
Defective or non-defective	95.2% accuracy	
Located surface	100% accuracy	
Width	0.58mm	1.99mm
Depth	0.23mm	0.64mm

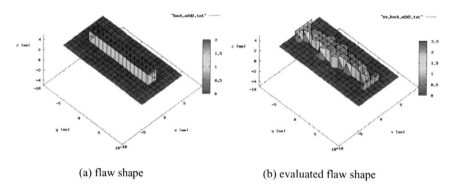

(a) flaw shape (b) evaluated flaw shape

Fig. 15. Evaluation of back surface flaw with 2mm width and 2mm depth

Table 4. Errors in evaluation of back surface flaw with 2mm width and 2mm depth

	Average error	Maximum error
Defective or non-defective	81.0% accuracy	
Located surface	100% accuracy	
Width	0.63mm	2.06mm
Depth	0.15mm	0.31mm

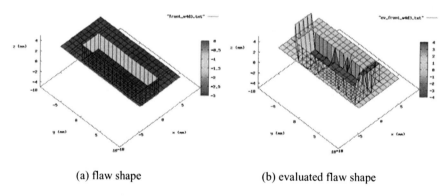

(a) flaw shape (b) evaluated flaw shape

Fig. 16. Evaluation of front surface flaw with 4mm width and 3mm depth

Table 5. Errors in evaluation of front surface flaw with 4mm width and 3mm depth

	Average error	Maximum error
Defective or non-defective	85.7% accuracy	
Located surface	93.3% accuracy	
Width	0.05mm	0.08mm
Depth	0.21mm	0.49mm

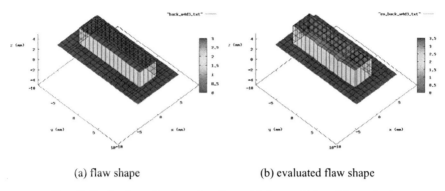

(a) flaw shape (b) evaluated flaw shape

Fig. 17. Evaluation of back surface flaw with 4mm width and 3mm depth

Table 6. Errors in evaluation of front surface flaw with 4mm width and 3mm depth

	Average error	Maximum error
Defective or non-defective	95.2% accuracy	
Located surface	100% accuracy	
Width	0.08mm	0.10mm
Depth	0.30mm	0.46mm

7 MFL Calculation of Oblique Flaw by FEM

Figure 18 shows the analytical model for the calculation of the MFL occurred near an oblique flaw. The specimen is magnetized by the magnetizing yoke. Distance between each magnetic pole is 115 mm. The material of magnetic core is silicon steel whose relative permeability is 20000. The specimen material is SS400 steel whose B-H curve is shown in figure 19. The length and the thickness of the specimen are 250 mm and 5 mm, respectively. And the specimen has an oblique flaw on the center of the specimen surface.

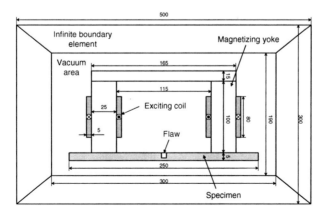

Fig. 18. Analytical model

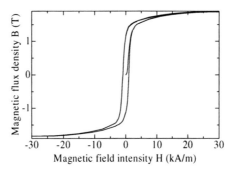

Fig. 19. B-H curve of SS400

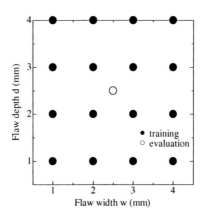

Fig. 20. Widths and depths of prepared oblique flaws

For training set of the NN, we introduced 16 flaws with width w and depth d indicated by solid circles in figure 20. We calculate the biaxial MFL distributions on the front and the back surfaces with respect to a magnetic sensor, and also in the case of the oblique angle θ is -60°, -30°, 0°, +30°, +60°, so that we obtain 160 training data. Meanwhile, for evaluation of unknown flaws, we prepare 10 front and back surface oblique flaws with width w and depth d given by a blank circle in figure 20.

8 Training of NN for Oblique Flaw

For the training of the NN to evaluate oblique flaw shapes, we use the 160 data calculated in the previous section. According to discussions in sections 2 and 3, we obtain all the parameters appearing in equations (11) and (12) for known oblique flaws.

First, we construct the localization-NN to judge the located surface of the flaw in the same way as in case of a parallelepiped flaw. In the localization-NN, the inputs are the ten CQs a_x, b_x, c_x, h_x, i_x, a_z, b_z, c_z, h_z, i_z and the outputs are the symbols f and b which represent the located surface, refer to figure 21.

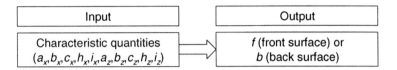

Fig. 21. Structure of localization-NN for evaluation of oblique flaw

Subsequently, we construct the identification-NN to identify the cross section of the flaw, i.e., the width, the depth and the oblique angle. In the identification-NN, the inputs are the above ten CQs and the symbol of the located surface f or b, and the outputs are the width, the depth and the oblique angle of the flaw, refer to figure 22. In this case, we add the located surface into the inputs of the NN for the improvement of the reconstruction accuracy.

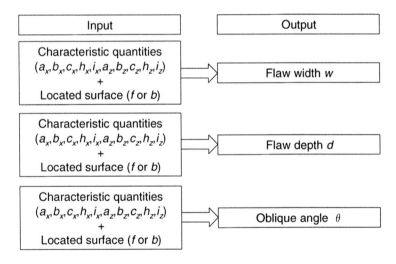

Fig. 22. Structure of identification-NN for evaluation of oblique flaw

9 Evaluation of Unknown Oblique Flaw

For the evaluation of unknown oblique flaws by the above trained NN, we use the data for 10 front and back surface oblique flaws calculated in section 7.

We evaluate the flaws in two steps. In the first step, we evaluate the located surface by the localization-NN. In the second step, we identify the width, the depth and the oblique angle of the flaw by use of the identification-NN. In this step, we use the evaluation results of the located surface in the first step as the input for the identification-NN. Besides, we adopt the average value of the d_x and d_z, which are obtained at the point of extraction of CQs by RAM, as the horizontal position of the flaw.

In figures 23 and 24, dashed lines show the flaw shapes and red lines show the evaluated flaw shapes. Table 7 shows the evaluation results for each unknown flaw and table 8 shows the averaged error and the maximum error in the evaluation.

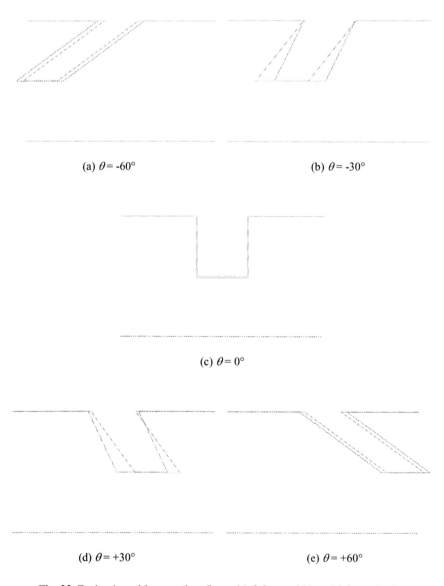

Fig. 23. Evaluation of front surface flaw with 2.5mm width and 2.5mm depth

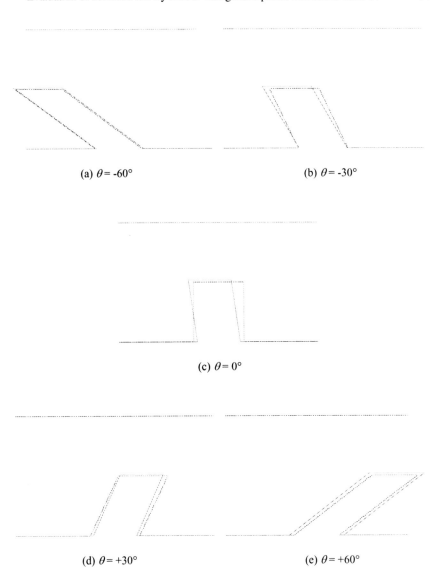

Fig. 24. Evaluation of back surface flaw with 2.5mm width and 2.5mm depth

In this instance, the located surface is estimated with 100% accuracy by the localization-NN. This result shows it is feasible to employ the evaluated located surface of the flaw as the input for the identification-NN. The averaged errors in evaluations of the width and the depth are around 0.05 ~ 0.31 mm. And the averaged error in evaluations of the oblique angle is lower than 5 degree. The each unknown flaw is evaluated with good accuracy.

Table 7. Evaluation results of unknown oblique flaws

Width	Depth	Located surface	Oblique angle	Estimated width	Estimated depth	Estimated located surface	Estimated oblique angle
2.5 mm	2.5 mm	Front	-60°	1.84 mm	2.48 mm	Front	-59.8°
			-30°	2.56 mm	2.51 mm		-42.6°
			0°	2.48 mm	2.55 mm		0.74°
			+30°	2.20 mm	2.52 mm		41.3°
			+60°	1.92 mm	2.51 mm		59.8°
		Back	-60°	2.63 mm	2.57 mm	Back	-59.8°
			-30°	2.43 mm	2.59 mm		-35.0°
			0°	2.11 mm	2.58 mm		-9.5°
			+30°	2.17 mm	2.57 mm		28.0°
			+60°	3.02 mm	2.58 mm		59.8°

Table 8. Errors in evaluation of unknown oblique flaws

	Average error	Maximum error
Located surface	100% accuracy	
Width	0.31 mm	0.66 mm
Depth	0.05 mm	0.09 mm
Oblique angle	4.19°	12.62°

10 Conclusion

In this paper, we attempt to evaluate the shapes of the surface flaw including its horizontal position and the located surface by biaxial MFLT using an amorphous MI sensor and a Neural Network (NN).

First, we evaluate three-dimensional shapes of a parallelepiped flaw in SS400 specimen.

(1) We adopt a method for evaluating three-dimensional shapes of a parallelepiped flaw, in which the three-dimensional flaw shapes are obtained by linking cross sectional evaluations on each measurement line.
(2) We proposed a new method for extracting the Characteristic Quantities (CQs) from the biaxial MFLT, i.e., Regression Analysis Method (RAM). We employ NN as the inverse analysis method to obtain the correlation between the above CQs and the flaw shape.
(3) It is found that the biaxial MFLT with NN using the CQs extracted by RAM can evaluate the three-dimensional shape and the location of the parallelepiped flaw with good accuracy.

Secondly, we evaluate two-dimensional shape of an oblique flaw whose MFL is calculated by FEM.

(1) We adopt the above proposed RAM for a method for extracting the CQs from the biaxial MFLT and employ NN for the inverse analysis.
(2) It is found that the biaxial MFLT with NN using the CQs extracted by RAM can evaluate the two-dimensional shape and the location of the oblique flaw with good accuracy.

References

[1] Jin, T., Que, P., Chen, T., Li, L.: Research on recognition algorithm of offshore oil pipeline defect inspection based on magnetic flux leakage method. J. JPI 48(5), 243–249 (2005)
[2] Mohri, K.: Application of amorphous magnetic wires to computer peripherals. Materials Science and Engineering A-Structural Materials Properties Microstructure and Processing 185(1-2), 141–146 (1994)
[3] Mohri, K.: Highly sensitive micro magnetic sensors for intelligent transport system (ITS). J. JSNDI 52(9), 468–472 (2003)
[4] Carvalho, A.A., Rebello, J.M.A., Sagrilo, L.V.S., Camerini, C.S., Miranda, I.V.J.: MFL signals and artificial neural networks applied to detection and classification of pipe weld defects. NDT&E International 39, 661–667 (2006)
[5] Han, W., Que, P.: 2D defect reconstruction from MFL signals by a genetic optimization algorythm. Russian Journal of Nondestructive Testing 41(12), 809–814 (2005)
[6] Hwang, K., Mandayam, S., Udpa, S.S., Udpa, L., Lord, W., Atzal, M.: Characterization of gas pipeline inspection signals using wavelet basis function neural networks. NDT&E International 33, 531–545 (2000)
[7] Joshi, A.: Wavelet transform and neural network based 3D defect characterization using magnetic flux leakage. International Journal of Applied Electromagnetics and Mechanics 28, 149–153 (2008)
[8] Abe, M., Biwa, S., Matsumoto, E.: 3D shape evaluation of rectangular flaw by biaxial MFLT with neural network. In: Proceedings of the 20th Symposium on Electro magnetics and Dynamics, pp. 45–50 (2008)
[9] Zatsepin, N.N., Shcherbinin, V.E.: Calculation of the magnetostatic field of surface defects. I. Field topography of defect models. Defektoskopiya 5, 50–59 (1966)
[10] Shcherbinin, V.E., Zatsepin, N.N.: Calculation of the magnetostatic field of surface defects. II. Experimental verification of the principal theoretical relationships. In: Defektoskopiya, vol. 5, pp. 59–65 (1966)
[11] Forster, F.: Nondestructive inspection by the method of magnetic leakage fields. Theoretical and experimental foundations of the detection of surface crack of finite and infinite depth. Soviet Journal of Nondestructive Testing 3(11), 841–859 (1982)

A Scalable Ultrasonic-Based Localization System Using the Phase Accordance Method

Toshio Ito[1], Tetsuya Sato[1], Kan Tulathimutte[1], Masanori Sugimoto[1], and Hiromichi Hashizume[2]

[1] Graduate School of Engineering, University of Tokyo, Tokyo, Japan
{toshio,sato,kantula,sugi}@itl.t.u-tokyo.ac.jp
[2] Information Systems Architecture Science Research Division, National Institute of Informatics, Tokyo, Japan
has@nii.ac.jp

Abstract. We introduce a localization system using ultrasonic signals. Our system adopts the Phase Accordance Method, which we proposed previously for accurate distance measurement [1, 2]. Unlike other ultrasonic-based systems, our system can locate ultrasonic transmitters with only one receiver, thanks to the high accuracy of the Phase Accordance Method. Thus our system can cover wider areas with fewer beacons than conventional systems, improving the scalability of the system. To evaluate the scalability and accuracy of our system, we developed a robot-tracking system and conducted an experiment with a moving robot. During the experiment, localization was executed both by our proposed method and by a conventional method based on trilateration. The experiment was repeated with different densities of receivers so that we could compare the accuracy and the scalability of our proposed method with those of the conventional method. We found that the position error of our method was degraded from 6.1 cm to 14.6 cm compared with the conventional method. However, our method improved the success rate of localization from 31% to 95%. We also conducted some localization experiments in which the robot stood still. This was because we wanted to investigate why the accuracy was degraded in the dynamic tracking. The results showed that the degradation of accuracy might be because of a systematic error in localization that is dependent on the geometric relationship between the transmitter and the receiver.

1 Introduction

Recently, the notion of location-aware computing has become very popular and has been the subject of much research. The essential part of location-aware applications is a localization system. The Global Positioning System (GPS) is one of the most widely known localization systems, but it cannot locate indoor objects. Various systems for indoor localization have therefore been proposed. In spite of the abundance of possible localization techniques, a decisively superior system that can be widely used has not yet emerged. One of the reasons for this is the trade-off between accuracy and the cost of deployment.

The goal of our research is to develop a low-cost, easy-to-deploy yet highly accurate localization system that can locate anything indoors. To obtain the required accuracy, we decided to adopt an ultrasonic method. Generally, there are two methods of ultrasonic localization, TOA (Time Of Arrival) and AOA (Angle Of Arrival). The common weakness of these two methods is that they can measure only one parameter (either distance or angle) per beacon, so they require three or more beacons spread out in the environment to locate the target uniquely. Because of this weakness, all of the ultrasonic-based systems proposed so far have required the beacons to be relatively densely deployed, which makes the systems expensive and deployment difficult.

To overcome this weakness of ultrasonic localization, we have proposed a new localization technology called the Phase Accordance Method [2]. The Phase Accordance Method precisely measures the time of flight of the ultrasound. Thanks to this method, the two-dimensional coordinates of the ultrasonic transmitter can be measured using only one receiver instead of three. Previous research into the Phase Accordance Method [1] has proved that this method performs very well for locating a fixed transmitter. However, we have not investigated the accuracy of the system when it tracks a target that moves constantly.

This paper is an extended version of our previous report [3], in which we conducted experiments with a moving robot to evaluate the accuracy and scalability of the proposed system. We tested the conventional TOA method as well for comparison. In the experiments, our system was able to track the moving robot continuously and showed sufficient accuracy. The conventional TOA method, on the other hand, was more accurate, but it tended to fail, especially when the receivers were sparse. These results demonstrated that with the Phase Accordance Method we can reduce the number of beacons to cover a given area, thus decreasing the cost and easing the deployment.

In addition to the experimental results described above, the present paper includes additional experiments in which we measured the position of a static robot. The results of these experiments helped us understand the cause of the accuracy degradation observed in the dynamic tracking experiments. The paper also includes a more detailed explanation of the Phase Accordance Method, our preliminary experiment [1] and our tracking system.

2 Related Work

Active Bat[4] is an ultrasonic-based localization system that uses the TOA method. In this system, an ultrasonic transmitter called the Bat is attached to each of the targets and ultrasonic receivers are deployed in the environment. Active Bat achieves a median error of 6 cm.

The Cricket location support system[5] addresses the privacy issue of the Active Bat by swapping the roles of the ultrasonic devices. This system was extended to the Cricket Compass[6], which is able to measure the orientation of the target as well as its position. In the Cricket Compass system, the receiver device has five ultrasonic sensors to determine its orientation. The localization error of the Cricket Compass was at best 5 cm. The Cricket's ability to track moving objects was investigated by Smith et al. [7]. They conducted localization experiments using a model train as a target. The median error of localization was at best 4 cm when the model train was moving at 1.43 m/s.

McCarthy et al. [8] proposed an algorithm to calibrate the beacons of their localization system automatically. With this algorithm, the beacons can take arbitrary positions. The standard deviation of the measured horizontal coordinates was reported to be 1.1 cm.

Hazas et al. [9] developed a localization system called Dolphin. The Dolphin transmitters modulate ultrasonic signals by Gold Codes, which makes the system more accurate and scalable. The localization error was reported to be within 2.5 cm.

All the above ultrasonic-based systems use the TOA method and so require many beacons in the environment. In Smith's experiment, for example, six beacons were used to cover the tracking area, which was about 3.5 m × 1.3 m. Unlike these systems, our proposed system requires only one receiver to estimate the 2D coordinates of the transmitter. Our receiver is able to locate the transmitter by accurately measuring the distance between them. To achieve such high accuracy in distance measurement, we have devised a novel method called the Phase Accordance Method, which can precisely detect the time of the arrival of the ultrasonic signal.

3 Ultrasonic Localization Methods

The most popular technique used in ultrasonic location systems is the TOA method. In this method the transmitter usually uses two channels, a radio-frequency (RF) channel and an ultrasonic channel. In the ultrasonic channel the transmitter sends out bursts of ultrasound, whereas in the RF channel the transmitter sends such information as the ID and coordinates as general digital data.

A receiver detects both the RF and ultrasonic signals and calculates the transmission delay time by comparing the arrival times of the two signals. Assuming that the RF packet reaches the receiver without delay, the transmission delay in the ultrasonic signal can be converted to the distance between the two nodes. When the receiver successfully estimates the individual distances of three transmitters with known coordinates, the receiver can calculate its spatial location by trilateration.

The main cause of errors in measurements in the TOA location system is the estimation error of the arrival time of ultrasonic signals. Ultrasonic transmitter and receiver devices resonate sharply at a specific frequency, e.g., 40 kHz for the most popular devices. Wave packets communicated via these resonators are deformed from their original shape, which makes determining the TOA at the receiver difficult.

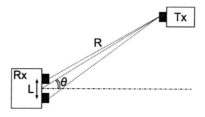

Fig. 1. AOA method

Another technology for ultrasonic localization is the AOA method. In systems using this method, receivers use two or more ultrasonic microphones in parallel, for estimating the positions of transmitters using triangulation.

Figure 1 shows the usual setting of the AOA method for 2D measurement. In this setting, two microphones are mounted on a receiver. The distance between the two microphones, L in Fig. 1, is called the baseline. The two microphones receive ultrasonic signals from the transmitter, and the phase difference between the two received signals, $\delta\phi$, is calculated. If the distance between the receivers and the transmitter $R \gg L$, the angle of arrival θ is estimated as:

$$\theta = \sin^{-1} \frac{c\delta\phi}{2\pi f L} \qquad (1)$$

where c is the speed of sound and f is the carrier frequency. However, there are difficulties in solving (1) because the phase ϕ is only retrieved from a sinusoidal carrier wave, which is a 2π-periodic function. If the observed phase difference is $\delta\phi$, the true phase difference may be $\delta\phi \pm 2\pi, \pm 4\pi, \ldots$. In most cases, multiple candidates for $\delta\phi$ give different, reasonable solutions of (1), hence the correct choice of θ is difficult.

To cope with the 2π-phase ambiguity, the Cricket Compass system, for example, employs three microphones for a planar measurement and five microphones for a stereo measurement to make simultaneous triangulations with different baselines.

Another difficulty in phase-oriented measurements of locations is the existence of multipath echoes. Objects around the ultrasonic wave channels reflect or scatter the sound wave, spoiling the phase accuracy.

4 The Phase Accordance Method

The Phase Accordance Method is a new technique for ultrasonic distance measurement. In this method, a burst of ultrasonic waves is used, similar to the previous TOA systems. The burst consists of subcarriers with two or more frequencies, and the phases of the carriers are in accord at a single time marker called the epoch. For simplicity's sake, let us consider a dual carrier system $f_1 + f_2$, e.g., $f_1 = 39.75$ kHz and $f_2 = 40.25$ kHz. The superposition of the two frequency signals makes a beat, which shows a distinctive pattern on

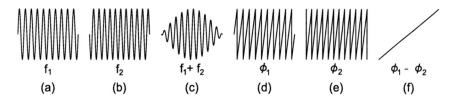

Fig. 2. Sync pattern. (**c**) shows a sync pattern, and (**a**) and (**b**) are its subcarriers. (**d**) and (**e**) show the phases of the subcarriers, and (**f**) is the difference between them.

the envelope (Fig. 2 (c)). The shape of the signal, however, does not matter in this method; instead, the phase difference of the component carriers is used. This beat pattern is called a sync pattern.

If the duration of a sync pattern is equal to the inverse of the frequency of the beat, i.e., $1/(f_2 - f_1) = 1/500 = 2$ ms in this case, the phase difference $\phi_2 - \phi_1$ sweeps from $-\pi$ to π, and the specific phase distance, e.g., $\phi_2 - \phi_1 = 0$, occurs only once in the pattern (Fig. 2 (f)). This point is used as the epoch, the time marker the method tries to find. The epoch can be determined very precisely by following the mathematical process described in the next section. This technique provides distance measurements within an error range of less than 1 mm under experimental conditions.

Another benefit of the Phase Accordance Method is its tolerance to multipath signals. As described above, the sync pattern is a burst signal with a duration of only 2 ms. This means that the sync pattern corresponds to a spatial length of about 70 cm. No multipath signals with paths 70 cm or longer than the real sync pattern can affect the localization.

4.1 Epoch Detection

When we describe the sync pattern as $s(t) = a_1 \sin(\omega_1 t + \phi_1) + a_2 \sin(\omega_2 t + \phi_2)$, we must know $\omega_1, \omega_2, \phi_1, \phi_2$ to detect the epoch. If the frequencies ω_1, ω_2 are given, then ϕ_1 and ϕ_2 can be calculated using the following mathematical process.

We define the inner product of two time domain functions $f(t)$ and $g(t)$ by:

$$\langle f(t), g(t) \rangle = \frac{1}{T} \int_{-T/2}^{T/2} f(t)\overline{g(t)} dt, \tag{2}$$

where $\overline{g(t)}$ is the complex conjugate of $g(t)$.

The inner product of $\sin(\omega t + \phi)$ and the complex exponential function $e^{j\Omega t} = \cos \Omega t + j \sin \Omega t$ yields:

$$\langle\sin(\omega t + \phi), e^{j\Omega t}\rangle = \frac{1}{T}\int_{-T/2}^{T/2}\sin(\omega t + \phi)\overline{e^{j\Omega t}}dt$$

$$= \frac{1}{2Tj}\int_{-T/2}^{T/2}\left(e^{j(\omega t+\phi)} - e^{-j(\omega t+\phi)}\right)e^{-j\Omega t}dt$$

$$= \frac{1}{2Tj}\int_{-T/2}^{T/2}\left(e^{j(\omega-\Omega)t+j\phi} - e^{-j(\omega+\Omega)t-j\phi}\right)dt$$

$$= \frac{1}{2j}\left(e^{j\phi}\operatorname{sinc}\frac{\omega-\Omega}{2}T - e^{-j\phi}\operatorname{sinc}\frac{\omega+\Omega}{2}T\right),$$

where $\operatorname{sinc} x = \sin x/x$ is the sampling function.

We can estimate the parameters of the sync pattern $s(t)$ by calculating the inner products of $s(t)$ using the $e^{j\omega_1 t}$ and $e^{j\omega_2 t}$ standard signals. Using the above formula, the inner product of $s(t)$ with $e^{j\omega_1 t}$ is:

$$2j\langle s(t), e^{j\omega_1 t}\rangle = a_1\left(e^{j\phi_1}\operatorname{sinc}\frac{\omega_1-\omega_1}{2}T - e^{-j\phi_1}\operatorname{sinc}\frac{\omega_1+\omega_1}{2}T\right)$$
$$+ a_2\left(e^{j\phi_2}\operatorname{sinc}\frac{\omega_2-\omega_1}{2}T - e^{-j\phi_2}\operatorname{sinc}\frac{\omega_2+\omega_1}{2}T\right).$$

Substituting $\operatorname{sinc}(-x) = \operatorname{sinc} x$ and $\operatorname{sinc} 0 = 1$:

$$2j\langle s(t), e^{j\omega_1 t}\rangle = a_1\left(e^{j\phi_1} - e^{-j\phi_1}\operatorname{sinc}\omega_1 T\right)$$
$$+ a_2\left(e^{j\phi_2}\operatorname{sinc}\frac{\omega_1-\omega_2}{2}T - e^{-j\phi_2}\operatorname{sinc}\frac{\omega_1+\omega_2}{2}T\right) \quad (3)$$

Similarly, the inner product of $s(t)$ and $e^{j\omega_2 t}$ is:

$$2j\langle s(t), e^{j\omega_2 t}\rangle = a_1\left(e^{j\phi_1}\operatorname{sinc}\frac{\omega_1-\omega_2}{2}T - e^{-j\phi_1}\operatorname{sinc}\frac{\omega_1+\omega_2}{2}T\right)$$
$$+ a_2\left(e^{j\phi_2} - e^{-j\phi_2}\operatorname{sinc}\omega_2 T\right). \quad (4)$$

Equations (3) and (4) can be combined into a matrix equation, where $a_1 e^{j\phi_1}$ and $a_2 e^{j\phi_2}$ are the unknowns:

$$2j\begin{pmatrix}\langle s(t), e^{j\omega_1 t}\rangle \\ \langle s(t), e^{j\omega_2 t}\rangle\end{pmatrix}$$
$$= \begin{pmatrix} 1 & \operatorname{sinc}((\omega_1-\omega_2)T/2) \\ \operatorname{sinc}((\omega_1-\omega_2)T/2) & 1 \end{pmatrix}\begin{pmatrix} a_1 e^{j\phi_1} \\ a_2 e^{j\phi_2}\end{pmatrix}$$
$$- \begin{pmatrix} \operatorname{sinc}\omega_1 T & \operatorname{sinc}((\omega_1+\omega_2)T/2) \\ \operatorname{sinc}((\omega_1+\omega_2)T/2) & \operatorname{sinc}\omega_2 T \end{pmatrix}\begin{pmatrix} a_1 e^{-j\phi_1} \\ a_2 e^{-j\phi_2}\end{pmatrix}. \quad (5)$$

Equation (5) can be strictly solved by examining the real part and the imaginary part individually. This means that we can obtain the phase and amplitude information about the components of the sync pattern by multiplying and integrating the received signal with $\overline{e^{j\omega_1 t}}$ and $\overline{e^{j\omega_2 t}}$.

After solving (5), we obtain the phases of the carriers at the center of the window, ϕ_1, ϕ_2. The epoch t_e is the time when the phases of the two carriers are in accord, so it is calculated as:

$$t_e = -\frac{\phi_1 - \phi_2}{\omega_1 - \omega_2} = -\frac{\phi_1 - \phi_2}{2\pi(f_1 - f_2)}. \tag{6}$$

Note that t_e is relative to the center of the window.

4.2 Integration of TOA and AOA

Because the Phase Accordance Method is a method for detecting the arrival time of the ultrasound, it can be easily used for TOA localization systems. Using the Phase Accordance Method, the receivers in the system can acquire more accurate distance estimates, and thus overall localization accuracy can be improved.

The accuracy of the Phase Accordance Method is so high that it can be applied to AOA systems as well. In AOA systems, two or more microphones are used and the differences between the phases of the received signals are usually measured, as described before. However, with the Phase Accordance Method the arrival time of the signal is so precisely determined that the angle can be calculated from the difference between the arrival times of the signals. In other words, the system no longer uses (1); instead, θ is calculated as:

$$\theta = \sin^{-1}\frac{c\delta t}{L} \tag{7}$$

where δt is the difference between the arrival times. Equation (7) gives a single, direct solution from a measurement whereas other phase-oriented methods using (1) often encounter the 2π-phase ambiguity problem.

As well as implementing the TOA and AOA methods individually, the Phase Accordance Method can also integrate the two methods into one. In the integrated method, ultrasonic receivers have two or more microphones, and transmitters send RF packets as well as ultrasound. After a receiver receives the sync pattern, the distances from each of the microphones to the transmitter are calculated using the Phase Accordance Method. Then the difference in the calculated distances is used to estimate the angle of the transmitter. Thus, using the integrated method, a single receiver can measure both the distance and the angle to the transmitter. This means that the coordinates of the transmitter can be determined by a single receiver and a single transmission of the sync pattern.

We conducted an experiment to investigate the performance of the integrated method in static localization [1]. In this experiment, an ultrasonic transmitter and a receiver were fixed on the floor so that their distance and angle (offset of the transmitter from the center direction of the receiver) formed about 3.2 m and 20°, respectively. This setting, however, was made by marking the positions on the floor with a conventional measuring tape,

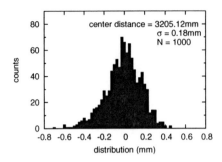 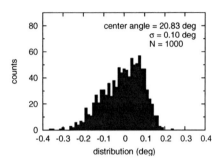

Fig. 3. Error distribution of distance measurements

Fig. 4. Error distribution of angle measurements

hence might be in error by a few millimeters. We performed the location measurements using the integrated method 1000 times, and recorded the distributions, which are shown in Figs. 3 and 4.

Readers may notice the amazing accuracy obtained by this method. For example, the standard deviation (which is the essential accuracy of the measurement – fixed errors can be eliminated by a compensation process if necessary) of the distance readings is only 0.18 mm, which is almost 1/100 of the previously reported errors of the TOA method. This accuracy is endorsed by a theoretical noise analysis [1].

We believe that the ability of a single receiver to locate transmitters lowers the density of the beacons that must be deployed, and the integrated method is obviously more tolerant to occlusions than conventional systems. These characteristics of scalability and robustness are very important, especially when deploying the system in a real environment. In addition to these advantages, it has been proved that the integrated method is able to locate fixed objects more accurately than conventional methods. However, to make the system more practical, it should be able to locate moving objects as well as static ones. Because locating moving objects is generally more difficult, the localization accuracy should be evaluated separately.

5 Experiment

To evaluate the ability of the integrated method to track a moving target, we developed a prototype system and conducted experiments with it. We will first introduce the prototype system and the experimental testbed, then describe the experiments and the results.

5.1 Prototype System

The prototype system we developed has three kinds of components, the transmitter, the receiver and the server PC. These components are equipped with

RF modules that enable them to communicate with each other using 2.4 GHz wireless signals. In our current system, one transmitter, one server and multiple receivers are used.

The transmitter is in charge of deciding when the ultrasonic signal should be transmitted and actually transmitting it. The transmitted signal is the dual carrier sync pattern described in Section 4; its central frequency is 40 kHz. Before the transmission, the transmitter broadcasts an RF packet to the whole system. We call this packet a trigger packet. Trigger packets are required so that the receivers can measure the propagation time of the sync pattern.

The receiver has two microphones, which are fixed so that the baseline $L = 80.2$ mm. The receiver begins to record incoming signals the instant it receives a trigger packet from the transmitter. The Phase Accordance Method is executed by a microcontroller, which then estimates the relative position of the transmitter. This is wirelessly sent to the server together with the unique ID of the receiver. We call this RF packet the location report.

While the receivers are receiving the sync pattern, the server waits for the location reports from the receivers. If multiple location reports are received, the server selects one and uses it to determine the absolute position of the transmitter. In our current system, the server picks up the location report with the smallest distance measurement, because it is likely to be the most reliable measurement. The server knows the absolute positions and orientations of the receivers. Thus, once the appropriate location report is selected, the server is able to locate the transmitter. If three or more location reports arrive at the server, it also performs the TOA method using the distance information in the reports. As a result, if sufficient location reports arrive, the server produces two localization results, one using the integrated method and one using the simple TOA method. We compared these two results in the experiment.

The ultrasonic speaker plays an important role in the system. The size of the area in which the system can locate the target mainly depends on the transmission characteristics of the speaker. To make the area larger, we adopted an omnidirectional speaker called US40KT-01, developed by Measurement Specialties, Inc. (Fig. 5). This is a cylindrical transmitter that can transmit 40 kHz ultrasonic signals in all directions.

We have examined the omnidirectivity of US40KT-01 by conducting a localization experiment. In this experiment, the speaker was fixed at a distance of 1 m from a receiver. Using the integrated method, the receiver located the speaker, which rotated step-by-step at the same position. Figures 6 and 7 show the distance and the angle measurements for each rotation angle of the speaker. The error bars represent the standard deviations of the measurements. From these figures we can see that all the measurements have very small deviations regardless of the rotation angle of the speaker. This means that US40KT-01 emits sync patterns in all directions without deforming them, allowing the system to perform stable localization.

Fig. 5. US40KT-01

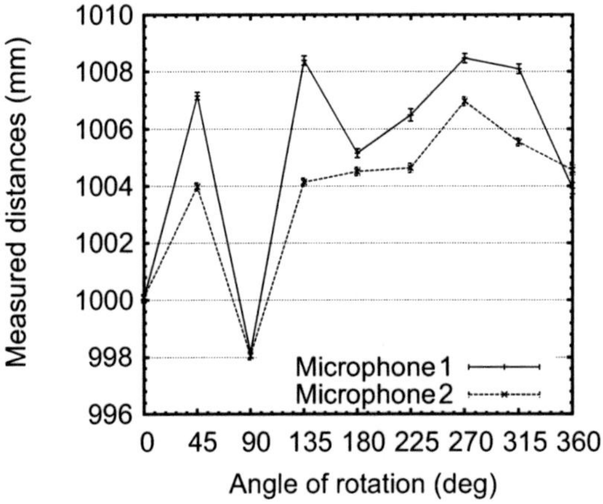

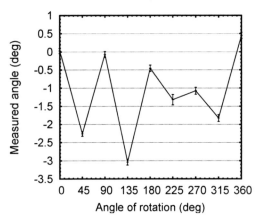

Fig. 6. Distance measurement using the omnidirectional speaker

Fig. 7. Angle measurement using the omnidirectional speaker

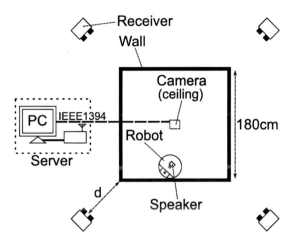

Fig. 8. Experimental testbed (top view)

5.2 Settings

We built up an experimental testbed in which the system tracks a moving robot. Figure 8 shows the geometrical setting of the testbed.

In this testbed, the objective of our localization system is to locate a robot moving within a tracking area surrounded by wooden walls. The transmitter is mounted on the robot and four receivers are fixed around the tracking area. The size of the tracking area is 180 cm × 180 cm. The receivers are placed d cm away from the tracking area, as seen in Fig. 8. To examine the relationship between the density of the receivers and the accuracy of the localization, we conducted the experiment three times, with $d = 125$ cm, 175 cm and 225 cm, respectively.

The robot used in this testbed is "iRobot Create", provided by the iRobot Corporation. iRobot Create is a disk-like robot with a diameter of about 34 cm. The robot is programmed to move in a straight line until it bumps into a wall. When it hits a wall, it rotates for a few seconds and then starts to go straight again. The velocity of the robot is about 140 mm/s.

The omnidirectional speaker US40KT-01 was attached to a metal bar that was mounted on the robot, so that the speaker was about 90 cm above the floor. This reduced the effect of multipath signals reflected from the floor. The receivers were mounted on tripods so that they were at the same height as the speaker.

One difficulty in evaluating our system is knowing the real position of the robot at the moment of localization. To achieve this, we decided to perform a vision-based localization synchronously with the ultrasonic localization. In the vision-based localization, a camera connected to the server PC captured an image of an infrared LED mounted on the speaker. The captured image was then processed by the PC to determine the actual coordinates of the robot from the position of the LED in the image. To make the calculation simpler, the camera was fitted with an infrared filter so that it could capture only the infrared LED of the robot.

To perform the vision-based localization accurately, the camera was attached to the ceiling, at a height of about 264 cm. In this setting, the length of 146 cm in the tracking area was mapped to about 940 pixels in the captured frame. Therefore, the effective resolution of this camera is about 0.155 cm/pixel. This means that the error of the vision-based localization is within a few millimeters.

The time synchronization between the vision and the ultrasonic localization was made by using the trigger packets. When the RF module on the server received a trigger packet, it ordered the PC to capture a frame from the camera immediately. This synchronization guarantees that the results of the two localizations refer to the robot at the same instant. Because the accuracy of the vision-based localization is much better than that of the ultrasonic-based one, we can consider the difference between the positions from the two localization systems as the error of our prototype system.

5.3 Procedure

Using the testbed described above, we conducted the following three kinds of experiments:

1. track the moving robot for 40 minutes;
2. locate the static robot at the center of the tracking area for 8 minutes;
3. locate the static robot at one corner of the tracking area for 8 minutes.

We changed the distance between the receivers and the tracking area (d in Fig. 8) in three steps: $d = 125$ cm, 175 cm, 225 cm. All the above experiments were conducted for each d; thus, we conducted nine experiments in total.

Localization was performed every 800 ms, so we obtained about 3000 localization results for the moving robot and 600 static results for each of the two experiments. One localization result consists of at most three position estimations: one using the vision system, one using the ultrasonic integrated method and one using the ultrasonic TOA method. Because of the attenuation of the ultrasound and environmental noise, the ultrasonic methods can fail to locate the transmitter.

The temperature in the room was 19.9°C.

5.4 Tracking a Moving Robot

First, we will consider the results of the experiments in which our system tracked the robot moving at a speed of about 140 mm/s. Note that the movement of the robot is independent of the localization system. It moves within the tracking area automatically as we described above.

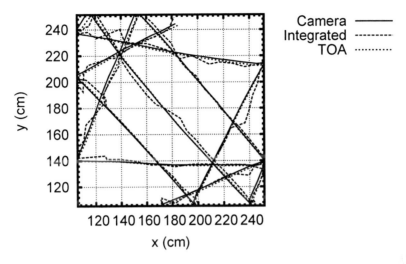

Fig. 9. Localized paths of the robot

Figure 9 shows part of the paths from the localization results acquired in the $d = 125$ cm setting. The figure indicates that both the integrated and the TOA methods were able to track the moving robot. Sometimes the localization results using the integrated method deviate from the true path by 10 cm or so, whereas the TOA method achieves stable tracking.

Figure 10 shows the cumulative distribution function (CDF) of the error in the measured positions of the robot. The figure includes the results of the two localization methods and the three settings of the testbed, as shown in the legend. The position error is defined as the distance between the position calculated using the vision-based system and the position given by the ultrasonic system. This is the case in the following context.

As we expected, the integrated method performs less accurately than the TOA method. The accuracy of the integrated method for $d = 125$ cm is more or less the same as the result for TOA with $d = 225$ cm. Not only is the accuracy of the integrated method relatively low as a whole, it is also more affected by the sparseness of the receivers. The 90th-percentile error moves from 6.18 cm to 14.62 cm when d is changed from 125 cm to 225 cm. However, an error of 14.62 cm is considered to be acceptable for distinguishing people in a room, which is the ability that most location-aware applications require.

Localization by the integrated method is based on two measurements, the distance measurement and the angle measurement, so the position error of the integrated method is a function of the errors in both measurements. The relationships between the three kinds of errors are described in Fig. 11. To scrutinize the accuracy of the integrated method in more detail, we calculated the three kinds of errors individually. Moreover, we classified these errors into

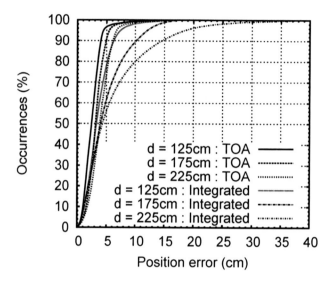

Fig. 10. Error CDF of the measured locations of the moving robot

five groups according to the distance between the transmitter and the receiver in charge of locating the robot. For each class, the standard deviations of the three errors are calculated, as shown in Fig. 12.

Figure 12 gives us important information. As the transmitter moves far away from the receiver, the position error increases rapidly. However, the distance error remains almost unchanged compared with the position error, while the angle error varies like the position error. Therefore, from Fig. 12, we can conclude that the deciding factor in the accuracy of the integrated method is the accuracy of the angle estimation.

In the integrated method, the angle is calculated by (7). Thus, as long as the accuracy of the distance measurement is the same, the accuracy of the angle estimation depends on L, the baseline of the receiver. In this experiment, we used receivers with a baseline of 80.2 mm. Apparently this baseline was so short that a slight degradation in the accuracy of the distance measurement produced a big error in the angle estimation. In a way, the TOA method had a much longer baseline, that is, the distance between the receivers. Using this long baseline, the TOA method was able to locate the transmitter more precisely.

Another criterion for evaluating the system is the success rate of localization. Intuitively, with sparse receivers, the system will be able to locate the target less frequently. The characteristic of this trade-off is one of the deciding factors of the scalability of the system.

Table 1 shows the success rates of the TOA and the integrated localization in the experiments. In this table, the TOA localization obviously tends to fail when the receivers are sparser, while the integrated method is almost always

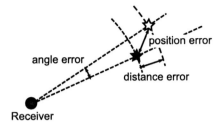

Fig. 11. Error definition. The white star represents the location acquired by the vision-based localization, while the black star represents the location from the ultrasonic-based localization.

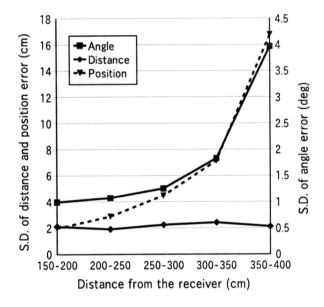

Fig. 12. Standard deviations (S.D.) of the three errors of the integrated method classified by the distance from the receiver that is in charge of localization

able to locate the robot. This is because the TOA method requires at least three location reports from the receivers. Mainly because of the attenuation of the ultrasonic signals, the probability for a single receiver to be able to locate the transmitter decreases as the distance between the receiver and the transmitter increases. The decline in this probability affects the TOA method much more than it does the integrated method.

To summarize the above results, the TOA method has the advantage of high accuracy, whereas the integrated method can cover a wider area. This result is as we had expected before conducting the experiments. Which method designers of location-aware applications should use depends entirely on the

Table 1. Success rate for locating the moving robot

d (cm)	Integrated (%)	TOA (%)
125	98.6	74.2
175	98.3	55.4
225	95.3	31.2

features of the application. If a designer requires the errors of localization to be always less than 5 cm or so, the TOA method should be used with many beacons. If the application does not require such high accuracy all the time, the designer can adopt the integrated method and lower the cost of deployment.

Of course, the two localization methods are not exclusive. If the integrated method is implemented, the TOA method can also be implemented easily, as we ourselves found. It is therefore possible for the application designers to vary the density of the receivers from place to place, considering the required accuracy and acceptable cost. This flexibility of the system surely expands the range of possible applications and lowers the overall cost of implementing them.

5.5 Locating a Static Robot

In this section, we will examine the results of the experiments in which the system locates the robot fixed at the center or near a corner of the tracking area. We conducted these experiments to determine the difference from the results of the dynamic experiments.

Figure 13 shows the position errors when locating the robot at the center. In this figure, the means of the position errors are plotted as points for each of the two localization methods and the values of d. The error bars in the figure represent the standard deviations of the position errors. At the $d = 225$ cm setting, the system did not receive enough location reports to perform the TOA localization, so the corresponding point is missing from the figure. Figure 14 is the same as Fig. 13 except that it shows the results when locating the robot near a corner.

In both figures, the standard deviations of the TOA method are less than 1 mm. The standard deviations of the integrated method are from 1.42 cm to 2.59 cm in Fig. 13, and from 0.27 cm to 0.98 cm in Fig. 14. The integrated method performs better when locating the robot in a corner than at the center. This is because the transmitter is nearer to the receiver. What is important here is that the standard deviations of the errors are significantly smaller as a whole than when the system tracks the moving robot. In Fig. 12, the standard deviation of the position error is at best 1.98 cm, which is nearly the worst case in Fig. 13. Obviously there is a difference between dynamic tracking and static localization.

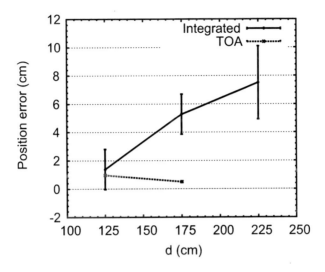

Fig. 13. Position errors when locating the robot at the center

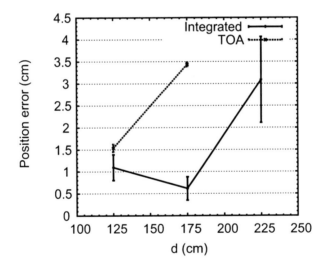

Fig. 14. Position errors when locating the robot near a corner

Another feature of these figures is the relatively large systematic errors. In Fig. 13, for example, the mean of the position errors in the integrated method increases as d becomes longer. Actually it increases beyond the range of the standard deviations. This behavior is more observable in the results of the TOA method, which have a much smaller deviation. The two figures thus suggest that the measurements of our localization system include some

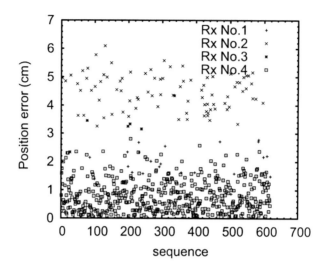

Fig. 15. Position errors of all the localization results of the integrated method in the experiment with $d = 125$ cm, with the robot at the center

systematic errors that are not negligible. In addition, the amount of the systematic errors differs between Figs. 13 and 14, and varies as d varies.

One possible cause of this varying systematic error is the directivity of the transmitter. Figure 6 shows that the systematic errors in the distance measurements vary as the speaker rotates. Although the range of the systematic errors is 1 cm or so, the systematic errors in the angle measurements vary on a larger scale, as seen in Fig. 7. The errors in the angle measurements affect the accuracy much more than the errors in the distance measurement, because they are multiplied by the distance. For example, a systematic error of 3° produces a position error of 15.7 cm if the distance is 3 m. As seen in Fig. 6, the behavior of the systematic errors in the distance measurements differs between the two microphones. This makes the systematic errors in the angle measurements fluctuate over such a wide range.

The relationship between the directivity of the transmitter and the systematic errors can also be seen in Fig. 15, which shows the position errors of all the localization results acquired by the integrated method when $d = 125$ cm and the robot is at the center. This figure also shows which receiver is responsible for each result. In this figure, receiver No. 4 accurately locates the robot, whereas receiver No. 2 has some systematic errors in its localization results. The receivers look at different sides of the transmitter, so this difference in systematic errors may be because of the directivity of the transmitter.

The results of the experiments show that there are some systematic errors in the location measurements. Furthermore, the preliminary experiment suggests that location measurement varies with the orientation of the transmitter. These results explain why the standard deviation of the localization in the dynamic experiment is larger than that obtained in the static experiment. In the dynamic experiment, the part of the transmitter from which the localizing receiver receives the signal constantly changes as the robot moves around; as well, the localizing receiver itself often changes. This makes the systematic error change constantly and results in the large random error observed in the previous section.

Systematic errors can generally be removed by calibration. In this case, however, it is impossible to cancel the error, because the receivers have no way of knowing the orientation of the transmitter. Given this knowledge and appropriate calibration, the accuracy of dynamic tracking could be improved to the extent that the standard deviation of the position errors was less than a few centimeters.

6 Conclusion and Future Work

We have developed a prototype localization system that implements the proposed integrated method. Using this system, we conducted robot-tracking experiments to investigate the accuracy and scalability of the integrated method and the conventional TOA method. We found that the 90th percentile of the position error was from 6.2 cm to 14.6 cm for the integrated method, and from 4.0 cm to 6.1 cm for the TOA method. As for scalability, the integrated method achieved continuous tracking with sparse beacons (four receivers in an area of about 5 m × 5 m), while the success rate of localization for the TOA method was only 31% in this configuration. From our results, we can conclude that the integrated method has better scalability at the cost of accuracy. We consider that appropriate combination of the integrated method and the TOA method broadens the applicability of ultrasonic-based localization systems. We also conducted static experiments in which the robot did not move. The dynamic and static experiments clarified issues that require further work.

First, we observed that the varying systematic errors in location measurements caused big errors in the tracking experiment. These systematic errors can theoretically be removed. However, to compensate for the errors we must modify our system so that it can measure the orientation of the robot as well as its position. One way to do this is to attach two or more transmitters to the robot. From the positions of the transmitters, the orientation of the robot can be calculated. The problem is that the present system can locate only one transmitter at a time, so we must devise a method that can intelligently estimate the orientation when the robot is moving.

Second, in the experiments, we used a robot that moved at a speed of about 140 mm/s. This is much slower than the speed of a walking person, which is about 1 m/s. To apply our system to location-aware applications, the system should be able to locate accurately targets that move at this speed. Because the Phase Accordance Method presumes that it knows the frequencies of the carrier waves, its accuracy is affected by the Doppler effect.

If the targets move faster than 1 m/s, as our preliminary experiments showed, the Doppler effect is no longer negligible and introduces large errors. We have already devised an algorithm to compensate for the Doppler effect; however, we have not yet tested it in realistic environments, mainly because of the lack of appropriate moving targets. We must conduct experiments to verify the ability of the system to track faster targets, so we can evaluate the system comprehensively.

References

1. Hashizume, H., Kaneko, A., Sugimoto, M.: Phase accordance method - an accurate ultrasonic positioning method and its characteristics. IEICE Transactions J91-A(4), 435–447 (2008)
2. Hashizume, H., Kaneko, A., Sugano, Y., Yatani, K., Sugimoto, M.: Fast and accurate positioning technique using ultrasonic phase accordance method. In: TENCON 2005, Melbourne, pp. 1–6 (2005)
3. Ito, T., Sato, T., Tulathimutte, K., Sugimoto, M., Hashizume, H.: A scalable tracking system using ultrasonic communication. In: 3rd International Conference on Sensing Technology (ICST 2008), Tainan, pp. 31–36 (2008), ISBN 978-1-4244-2176-3
4. Ward, A., Jones, A., Hopper, A.: A new location technique for the active office. IEEE Personal Communications 4(5), 42–47 (1997)
5. Priyantha, N.B., Chakraborty, A., Balakrishnan, H.: The cricket location-support system. In: MobiCom 2000: Proceedings of the 6th Annual International Conference on Mobile Computing and Networking, Boston, pp. 32–43 (2000)
6. Priyantha, N.B., Miu, A.K.L., Balakrishnan, H., Teller, S.: The cricket compass for context-aware mobile applications. In: MobiCom 2001: Proceedings of the 7th Annual International Conference on Mobile Computing and Networking, Rome, pp. 1–14 (2001)
7. Smith, A., Balakrishnan, H., Goraczko, M., Priyantha, N.: Tracking moving devices with the cricket location system. In: MobiSys 2004: Proceedings of the 2nd International Conference on Mobile Systems, Applications, and Services, Boston, pp. 190–202 (2004)
8. McCarthy, M., Duff, P., Muller, H.L., Randell, C.: Accessible ultrasonic positioning. IEEE Pervasive Computing 5(4), 86–93 (2006)
9. Hazas, M., Hopper, A.: Broadband ultrasonic location systems for improved indoor positioning. IEEE Transactions on Mobile Computing 5(5), 536–547 (2006)

VHDL-AMS Modelling of Ultrasound Measurement System in Linear Domain

D. Kourtiche, R. Guelaz, A. Rouane, and M. Nadi

Electronic Instrumentation Laboratory of Nancy
Nancy University, France

Abstract. Piezoelectric materials are commonly used in many applications. Different approaches were developed to predict the piezoelectric transducer behaviour. Among them, the resolution of piezoelectric equations by numerical methods is currently used. Another method is based on the equivalent electrical circuit simulation : Pspice or VHDL–AMS tools. This article proposes VHDL-AMS model for a pulse echo ultrasonic system. The simulation is based on the Redwood model and its parameters are deduced from the transducer acoustical characteristics. The electrical behaviour of the proposed model is in very good agreement with the real system behaviour.

1 Introduction

Ultrasonic transducers are a key element that govern the performances of both generating and receiving ultrasound in an ultrasonic measurement system. Prediction of ultrasonic systems behaviour needs simulation tools adapted to transducers methodology conception. The design and optimisation involve a global knowledge of interactions between acoustic, electronics and the ultrasonic properties of the propagation medium.

To modelize piezoelectric ceramics, is not so easy because they interface with mechanical and electrical domains. To design an ultrasound system, it is essential to know precisely its behaviour in these two domains according to the propagation medium.

The electromechanical interaction, represented by electrical equivalent circuits, was first introduced by Mason. He have shown that for one-dimensional analysis most of the difficulties in deriving the solutions could be surmounted by borrowing from electrical network theory. He proposed an exact equivalent circuit that separated the piezoelectric material into an electrical port and two acoustical ports through the use of an ideal electromechanical transformer [1,2]. The problems with the model is that it required a negative capacitance at the electrical port.

Redwood [3] improved this electromechanical model by incorporating a transmission line, making possible to extract useful information on the temporal response of the piezoelectric component.

Redwood's model was simulated by Morris and Hutchens [4] with SPICE software. An alternative equivalent circuit : the KLM model was presented by Krimholtz, Leedom and Matthai [5]. This analogous circuit consists on a frequency-dependent network. It includes a transformer connected to the middle of a transmission line. This model has been widely used in the medical imaging to design high frequency transducers. This design is advantageous in applications such as acoustically cascaded ones.

Trying to remove the negative capacitor, as it is in Mason's model, and the frequency dependent transformer as it is in the KLM model, Leach [6] proposed a model with controlled current and voltage sources replacing the transformer. To include acoustical attenuation effect, Püttmer [7] used a lossy transmission line in Leach's model. Several ultrasound systems was implemented into SPICE for simulation [8-10]. In this context, the simplicity with which the conditioning electronics can be designed using SPICE makes it important to implement the theoretical model of the transducer with a similar software tool. In addition, with SPICE it is easier to evaluate the performance of the transducer, both in time and frequency domains. Limitations are in the restrictive components library which could not be easily extended for implementation of complex equation like differential equations.

Ultrasound systems are widely used. They find many applications in engineering, medicine, biology, and other areas [11,12]. Modelling and simulation of such systems is a difficult task due to the presence of multi-physics effects and their interactions. VHDL-AMS language is appropriate for ultrasonic systems methodology conception because it could take into account all the transducer environment including microelectronic stimulation and acoustic load. The use of behavioural models in simulation simplify physics and explore interactions between different domains in a reasonable amount of time [13,14].

This paper presents a method for multi-domain behavioural modelling of ultrasound measurement system. We validated this methodology through a study cases in linear ultrasound measurement in which the ultrasound transducer model takes great importance. To perform the implementation, a virtual-prototyping environment, ADVance® MS (ADMS) tool from Mentor Graphics is used. This environment provides multi-level model integration required for real systems design and analysis.

Section two describes how to implement a VHDL-AMS model for the transducer vibrating in the thickness mode. We based our study on Reedwood model, which gives an analogy between physical behaviour of piezoelectrical material and the implementation with basic electric elements into an electrical diagram. The third section gives a description of pulse-echo measurement system with VHDL AMS. The medium of propagation is modelled by a transmission line. We studied the influence of the backing and the propagating mediums on the transducer response. The results of simulation presented in section four gives the transducer frequency response and modelling of the ultrasound pulse echo system behaviour in the linear domain.

The fifth section resumes experimental measurement results done in order to test the validity of the proposed model. The experimental setup consists on an ultrasonic transducer operating in pulse-echo mode and excited by a PCB (Matec®*TB1000* card) that delivers burst waves. The transducer is plunged in the medium (liquid) under investigation and radiate at a central frequency of 2.25 MHz to a perfect reflector which reflects waves to the same transducer.

Using this setup, we have measured the transducer sensitivity versus the amplitude and the frequency of the excitation. Variations of acoustic velocity in some mediums according to temperature have been measured besides. Section six concludes this contribution and its perspectives.

2 The Transducer Model

Ultrasonic transducer plays a key role in ultrasonic measurement system as it generates and receives the ultrasonic waves. Usually this complex electro-mechanical device is difficult to characterize. For an optimal use of the transducer, it is important to know its transmitting and receiving properties. These properties are fundamental to design the transducer in the form of an equivalent electrical circuit [1-4].

The implementation of this equivalent electrical circuit in a simulation tool such as PSpice allows the electronic integration of piezoelectric devices [15,16].

To design ultrasonic transducers, many authors find it convenient to use a three port model. Models of this type that are commonly used in practice are: the Mason model, the Redwood model and the KLM model [3,6]. Both models treat the transducer as a ceramic piezoelectric element where both one dimensional electrical and mechanical fields are present (figure 1).

The electrical port corresponds to the electrical connections with the plated faces of the piezoelectric ceramic while the two acoustic ports are the two faces of the ceramic.

From the equation of acoustic wave's propagation in piezoelectric materials, it is possible to write linear relations that link the mechanical magnitudes (force F and speed of particles v) which are preserved at an interface to electrical quantities (applied potential V_3 and intensity I_3 of the current).

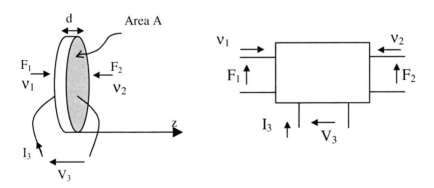

Fig. 1. Model (1-D) for electrical and acoustical parameters for a piezoelectric ceramic and its representation as a three port system

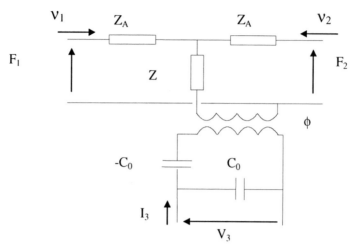

Fig. 2. Mason's equivalent circuit model of three port system defined by matrix 3x3 $Z_A = -j\, Z_t \tan(k\,d/)$, $Z = -j\, Z_t \sin(k\,d)$, $\phi = h_{33}\, C_0$

The electrical and mechanical lumped parameters for this three port model correspond to the following 3x3 electromechanical matrix.

$$\begin{bmatrix} F_1 \\ F_2 \\ V_3 \end{bmatrix} = -j \begin{bmatrix} Z_t \cot(kd) & Z_t/\sin(kd) & \dfrac{h_{33}}{\omega} \\ Z_t/\sin(kd) & Z_t \cot(kd) & \dfrac{h_{33}}{\omega} \\ \dfrac{h_{33}}{\omega} & \dfrac{h_{33}}{\omega} & \dfrac{1}{\omega C_0} \end{bmatrix} \begin{bmatrix} v_1 \\ v_2 \\ I_3 \end{bmatrix} \qquad (1)$$

The multiple parameters appearing in this model are defined below :

v_1 and v_2 (m/s) are the acoustic particles velocities at the front and the back faces of the disk, the parameter k is the wave number for the piezoelectric ceramic, $k = \omega/v^D$, where ω the angular frequency (rad/s), v^D (m/s) is the wave speed of compressional waves in the piezoelectric plate given by $v^D = \sqrt{C_{33}^D/\rho}$ in terms of the elastic constant of the piezelectric ceramic, C_{33}^D (N/m^2) , at constant electric flux density, and ρ (kg/m^2), the density of the ceramic.

h_{33} (V/m) , is the piezoelectric stiffness constant for the ceramic, and C_0, the clamped capacitance of the plate. C_0 is given by $C_0 = A/\beta_{33}^s d$ where A (m^2) is the area of the ceramic and β_{33}^s (m/F) is the dielectric impermeability of the ceramic at constant strain, and d (m) is the ceramic thickness.

The quantity $Z_t = A\,\rho\,v^D$ (Rayl) is the plane wave acoustic impedance of the piezoelectric ceramic, while Z_B (Rayl) is the corresponding acoustic impedance of the backing, which is a function of frequency.

Figure 2 gives the Mason equivalent circuit model. In this model, the electrical port is coupled, through an ideal transformer (transformation ratio $\phi = h_{33}\,C_0$) to a T-network which represents the acoustic propagation in the piezoelectric material. The mechanical ports of the transducer are located at the extremes of the T-network. The presence of the negative capacitance $-C_0$ close to the electrical terminals is to be noted. Both, the negative capacitance and the ideal transformer participate to govern the electromechanical coupling model.

This circuit correspond exactly to the 3x3 electromechanical matrix (equation 1) and can be deduced from them. Their complete equivalence can be established from the equality of the open and short circuit input electrical impedance.

3 Ultrasound Pulse-Echo System

The most common configuration of an ultrasonic system widely used for acoustical measurements is shown on the figure 3.

It involves the generation, propagation and reception of the signal. The system operates in pulse echo mode. The ultrasonic waves generated by the transducer propagate through the medium and the received echo is converted by the same transducer to electrical signal.

The ultrasonic transducer is a bi-physical device that transforms electrical signal to acoustical wave and *vice versa*.

To obtain a VHDL-AMS description of this pulse echo measurement system, one must give:

i)- a description of the ultrasonic transducer by using equivalent circuit of Mason as adapted by Redwood. The mechanical part of the piezoelectric transducer is easily represented using a transmission line model, two parameters are sufficient to entirely define the mechanical part of the transducer, the impedance and the sound propagation delay through the transducer.

ii)- a description of the propagation medium by means of a transmission line. This model corresponds to the electric equivalent circuit of Branin [17]. The transmission line parameters are calculated according to the characteristics of the medium.

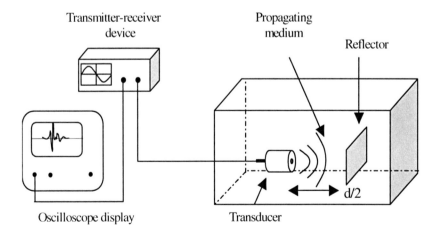

Fig. 3. Bloc diagram of experimental setup

3.1 VHDL-AMS Behavioural Model of Transducer

The Redwood equivalent circuit shown in figure 4 is similar to Mason circuit. It can be obtained from Mason model by substituting the *T* network by the equivalent transmission line. This model has been used for Spice implementation of the transducer. In our study this model is implemented with the VHDL-AMS behavioural language.

The model is divided in two parts. The first one, is the electrical port which includes the capacitors C_0 and $-C_0$ that represent the capacitance motional effect. This electrical port is connected to a resistance R and a voltage source noted V_3. The second part is composed by the two acoustic ports, T is an ideal electro-acoustic transformer with a ratio $h_{33}.C_0$.

Piezoceramic layer is assimilated to a propagation line characterized by its thickness d, characteristic impedance Z_t, and the c_0 the acoustic velocity.

One branch of the piezoceramic layer is in contact with the back medium (Z_{back}) and the other is in contact with the propagation medium (Z_{front}).

Transducer modelling with VHDL-AMS language (figure 5) is based on writing of the different equations of the Redwood's model elements.

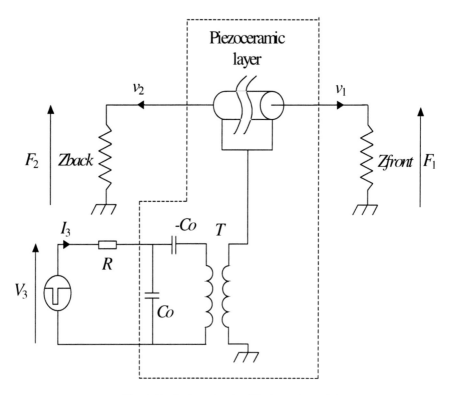

Fig. 4. Equivalent circuit of Redwood's model

A VHDL-AMS model is a source code where physical behavioural of an element is described with a specified semantic referenced in a IEEE Standard [18].

In this approach, electrical and acoustic elements are described. Electrical element is defined with voltage and current quantities and acoustic element with force and velocity quantities. Libraries should be called before elements description as illustrated below :

```
LIBRARY disciplines;
USE disciplines.electromagnetic_system.all;
USE disciplines.kinematic_system.all;
```

The VHDL-AMS implementation of Redwood's equivalent model (Figure 4) is given by:

```
ENTITY Redwood IS
GENERIC (C0,kt,Zt,Td : real);
PORT (TERMINAL p, m : electrical;
TERMINAL t11, m11, t22, m22: kinematic_v);
END ENTITY Redwood;

ARCHITECTURE structure OF Redwood IS
terminal p1 : electrical;
terminal t1,t1x,t2x : kinematic_v;
QUANTITY v1 across i1 through p TO m;
QUANTITY v2 across i2 through p TO p1;
QUANTITY vte across ite through p1 TO m;
QUANTITY pti across uti through t1 TO kinematic_v_ground;
QUANTITY p1xr across u1xr through t1x TO t1;
QUANTITY p2xr across u2xr through t2x TO t1;
QUANTITY p1x across u2x through t1x TO t11;
QUANTITY p2x across u1x through t2x TO t22;
QUANTITY p11 across     t11 TO t1;
QUANTITY p22 across     t22 TO t1;

BEGIN

i1 == C0 * v1'dot;
i2 == -C0 * v2'dot;
pti == kt * vte;
uti == -ite/kt;
p1xr == p22'DELAYED(Td) - p1x;
p2xr == p11'DELAYED(Td) - p2x;
p1x == (u1x+u2x'DELAYED(Td))*Zt/2.0;
p2x == (u2x+u1x'DELAYED(Td))*Zt/2.0;

END ARCHITECTURE structure;
```

Fig. 5. VHDL-AMS code of the Redwood model

3.2 Linear Propagation Medium Modelling

Different non piezoelectric layers, matching layers or propagating medium, can also be represented by transmission lines. Electrical network analysis can then be used. The linear propagation with VHDL AMS language is assimilated to an electric line propagation behaviour without losses.

The model suggested below is thus consider as an acoustic propagation medium in which an incident ultrasonic wave is delayed with a T_d time corresponding to the flight time of the acoustic wave in material. This flight time is related to the relation $T_d = c_0.d$ with c_0 the characteristic propagation celerity of the medium and d the

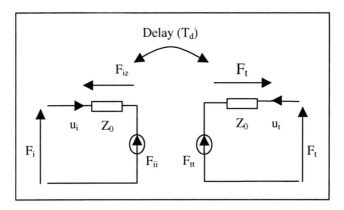

Fig. 6. Equivalent electric diagram of the acoustic linear propagation in a medium

```
ENTITY LinearMedium IS
GENERIC (Z0,Td : REAL);
PORT (TERMINAL p1,m1,p2,m2 : Kinematic_v);
END LinearMedium;

ARCHITECTURE structure OF AcousticLayer IS

TERMINAL t11, t22 : Kinematic_v;
QUANTITY Fi ACROSS p1 TO m1;
QUANTITY Ft ACROSS p2 TO m2;
QUANTITY Fii ACROSS uiz THROUGH t11 TO m1;
QUANTITY Fiz ACROSS ui THROUGH t11 TO p1;
QUANTITY Ftz ACROSS ut THROUGH t22 TO p2;
QUANTITY Ftt ACROSS utz THROUGH t22 TO m2;

BEGIN
Ftt == Fi'DELAYED(Td) - Ftz;
Fii == Ft'DELAYED(Td) - Fiz;
Fiz == (uiz + utz'DELAYED(Td))*Z0/2.0;
Ftz == (utz + uiz'DELAYED(Td))*Z0/2.0;

END ARCHITECTURE structure;
```

Fig. 7. VHDL-AMS code of the linear propagation medium

thickness. Z_0 is the characteristic acoustic impedance. This model corresponds to the electric equivalent circuit of Branin [13, 14, 17] presented in figure 6.

Indices i and t referred respectively to the incident part and transmitted part of the transmission line.

Notation u refers to velocities in the medium, F to Forces. F_{ii}, F_{iz}, F_{tt} and F_{tz} are defined in the VHDL-AMS model by including the delay T_d. This propagation line model is assimilated to a medium layer submitted to an incident wave. It is modeled by the following algorithm us presented in figure 7.

The acoustic source amplitude is normalized to be 1 Pa in order to compare the behaviours of various transducers under the same conditions.

3.3 VHDL-AMS Modelling of the Experimental Setup

The global schema with the pulse echo transducer implemented with Redwood's model is presented figure 8.

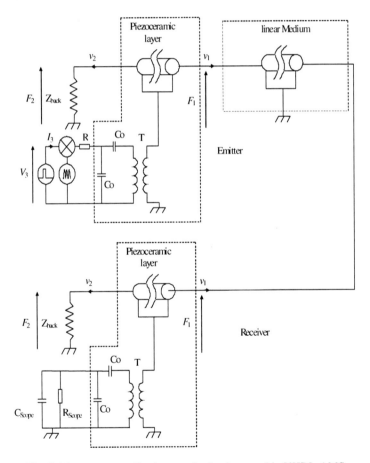

Fig. 8. Measurement cell schema to be implemented in VHDL-AMS

```
ENTITY Measure_cell IS
END Measure_cell;

ARCHITECTURE structure OF Measure_cell IS

TERMINAL n1,n2,n3,n4,n5,n6,n8,n9 : ELECTRICAL;
TERMINAL Tb,Tb2,Tf,Tf2 : kinematic_v;
CONSTANT A : real := 132.73e-3
CONSTANT e: real := 1.0e-3;
CONSTANT Co: real := 1109.8e-12;
CONSTANT Va : real := 4530.0;
CONSTANT kt : real := 0.49;
CONSTANT epsi0 :real:= 8.8542e-12;
CONSTANT epsi33 :real:= 650.0;
CONSTANT ro :real:= 3300.0;
CONSTANT h :real:=   kt*Va*sqrt(ro/(epsi0*epsi33));
CONSTANT K : real := h*Co
CONSTANT ZT : real := 34.9e6;
CONSTANT Zfront : real := 1.5e6;
CONSTANT Zback : real := 445.0;
QUANTITY   vinput   across   ie   through   n1   to
electrical_ground;

BEGIN

If now > 0.0 and now < 222.2e-9 USE
vinput == -100.0;
Else vinput == 0.0;
End USE;

R1 : entity Resistor(bhv) generic map (50.0)
port map ( n1, n2);
T1 : entity Redwood(bhv)  generic map( Co, K, A*ZT,
e/Va)
port map( n3, kinematic_v_ground,
n4, kinematic_v_ground, n2, electrical_ground);
Medium : entity linearMedium(bhv)
generic map (1.5e6, 20e-9 ,fo ,1500.0 ,5.0 ,0.9 ,0.045
,1.0e-9)
port    map(    n3,    kinematic_v_ground,    n5,
kinematic_v_ground);
back : entity Resiskinematic(bhv)
generic   map   (   A*Zback   )   port   map   (   n4,
kinematic_v_ground );
back2 : entity Resiskinematic(bhv) generic map (
A*Zback )
port map ( n8, kinematic_v_ground );
T1 : entity Redwood(bhv) generic map( Co, Kt, A*ZT,
e/Va)
port map( n3, kinematic_v_ground, n8,
kinematic_v_ground, n9, electrical_ground );
RScope : entity Resistor (bhv) generic map ( 1.0e6)
port map ( n9, ground);
Cscope : entity Capacitor (bhv) generic map ( 13.0 e-
12) port map ( n9, ground);

END ARCHITECTURE structure;
```

Fig. 9. VHDL-AMS code of the measurement cell

The inferior part is the transducer in reception mode and its connected to a charge (C_{scope}, R_{scope}) which represent the input impedance of the electric measurement tool.

The piezoceramic layer equivalent circuit is shown as a linear propagation diagram.

The superior part correspond to the transducer in emission mode which is connected to the electric source. The associated test-bench with VHDL-AMS language of the experimental setup is presented in figure 9.

4 Simulation

The measurement cell modelling implemented with VHDL-AMS presented on the figure 8 is simulated. The amplitude of the transducer is fixed at 1Volt with a frequency of 2.25MHz.

The time sampling step used for discretization is $1.10^{-9}s$. Mediums analyzed in simulation are compared with the measurement cell results. Table 1 gives acoustic characteristics of simulated mediums.

Table 1. Acoustic Characteristic Of Mediums used in simulation [19]

Medium	Acoustic Impedance MRayls	Acoustic velocity m/s
Water at 25° C	1.494	1469
Ethanol at 25° C	0.91	1158
Paraffin oil	1,86	1420
Sea water at 25°C	1,569	1531

4.1 Transducers Frequency Response

The frequential transducers response study is essential to predict the sensitivity of the system for the various analyzed mediums.

The transducers were built with *PZT* ceramic of *P188* type (*Quartz et Silice®*) witch characteristics are summarized on table 2.

Table 2. Transducers Acoustic Characteristics

Parameters	Quantity	Value
F_0	Frequency resonance	2.25 MHz
A	Area	132.73 mm²
d	Thickness	1 mm
Z_t	Acoustic impedance	34.9 MRayls
c_0	Acoustic velocity	4 530 m/s
C_0	Capacitor of the ceramic disc	1 109.8 pf
E_{33}	Dielectric constant	650.0
k_t	Thickness coupling factor	0.49
h_{33}	Piezoelectric Constant	$1.49.10^{+9}$

The tool used for VHDL-AMS simulation is *ADMS v3.0.2.1* of Mentor Graphics Company. To perform the transducer frequency analysis, we used a VHDL-AMS *Testbench* where a frequential source is connected to the electrical input.

The simulation code used to obtain the transducer electrical impedance characterization is presented in figure 10.

```
ENTITY Impedance_Simulation IS
END Impedance_Simulation;
ARCHITECTURE struct OF Impedance_Simulation IS

  TERMINAL n1,n2 : ELECTRICAL;
  TERMINAL n3,n4 : KINEMATIC_V;
  CONSTANT Va :real:= 4530.0;
  CONSTANT ZT:real:= 34.9e6;
  CONSTANT e :real:= 1.0e-3
  CONSTANT A :real:= 132.73e-3
  CONSTANT fo :real:= 2.25e6
  CONSTANT kt :real:= 0.49;
  CONSTANT epsi0 :real:= 8.8542e-12;
  CONSTANT epsi33 :real:= 650.0;
  CONSTANT ro :real:= 3300.0;
  CONSTANT h :real:= kt*Va*sqrt(ro/(epsi0*epsi33));
  CONSTANT Co :real:= A*epsi33/e;
  QUANTITY vac : real spectrum 1.0,0.0;
  QUANTITY vinput across ie through n1 to
  electrical_ground;

BEGIN

  Vinput == vac;
  Rgen : ENTITY resistanc   (bhv)  GENERIC MAP (50.0)
  PORT MAP (n1, n2);
  T1 :  entity Redwood(bhv) generic map( Co, K, A*ZT, e/Va)
  PORT MAP (n3, kinematic_v_ground,    n4, kinematic_v_ground, n2,
  electrical_ground );
  Rfront : ENTITY resistanc (bhv) GENERIC MAP (1.5e6*A)
  PORT MAP (n4, ground);
  Rback : ENTITY resistanc (a) GENERIC MAP (A*445.0)
  PORT MAP (n3, electrical_ref);

END ARCHITECTURE struct;
```

Fig. 10. VHDL-AMS code of the transducer impedance simulation

Figure 11 shows respectively the results obtained with the acoustic value of the transducer. Simulations were performed for water and ethanol.

This result shows a resonance frequency at 2.26 *MHz* that corresponds typically to the frequency behaviour of an ultrasonic transducer input impedance vibrating at its fundamental characteristic.

4.2 Measurement Cell Simulation

One first result, was to compute the transducer sensitivity which is defined as the ratio of the echo's amplitude to the excitation one.

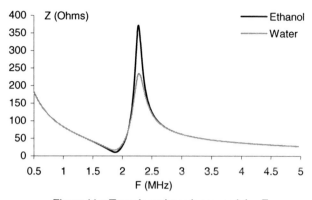

Figure 11a: Transducer impedance modulus Z

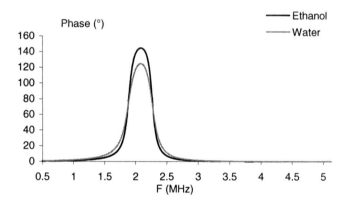

Figure 11b. Transducer impedance phase

Fig. 11. Transducer impedance modulus Z (a) and phase (b) of the emitter vibrating at 2.25MHz

The figure 12 gives the sensitivity of transducer versus the excitation frequency with air as backing medium and water as medium of propagation.

Thus, one can study the effect of backing medium on the band width and the sensitivity. To derive the value of the sensitivity we represent the signal V_{echo} versus supply voltage V_{in} (figure 13). We obtain $\xi = 0.032$ for the sensitivity.

In order to test the validity of our model, we use it to determine the acoustic velocity of some common and well known materials such as distilled water, ethanol, paraffin oil and sea water. Thus, the sound speed is determined by using the well known expression :

$$c = \frac{d}{\tau} \qquad (2)$$

τ is the time spent by the wave while traveling through the medium. The width of the medium is fixed while τ is determined from the system transient response.

In figure 14, the transient response in distilled water is shown, while figure 15 gives it for paraffin oil as propagation medium.

Using this method, we find the following results:

For water c_{water} = 1480 m/s and for paraffin oil $c_{paraffin}$ = 1520 m/s. These values of sound speed are in concordance with those published in the literature [19].

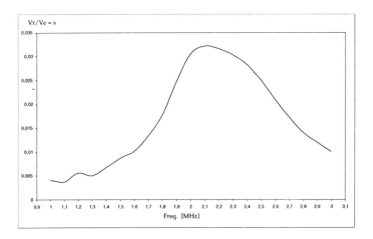

Fig. 12. The transducer's sensitivity versus excitation frequency

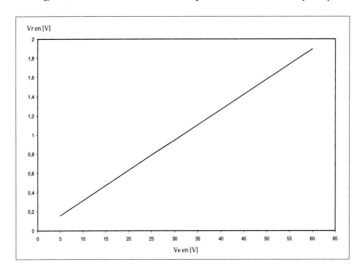

Fig. 13. The transducer's sensitivity versus excitation amplitude

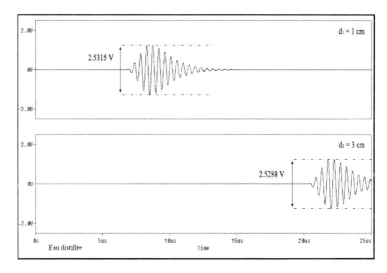

Fig. 14. Simulation of a pulse propagation in distilled water

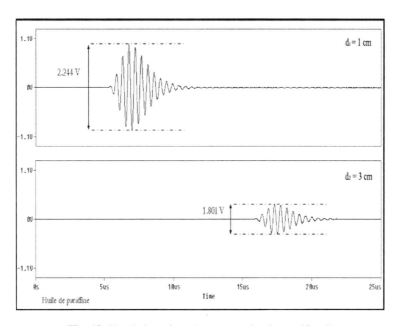

Fig. 15. Simulation of a pulse propagation in paraffin oil

5 Experimental Results

In order to excite the transducer, the Matec® *TB1000* card was used [20]. Incorporated in a personal computer and driven by a *C++* software program, the

TB1000 card allows to generate a powerful sinusoidal tone burst pulser/receiver with variable amplitude, frequency, width and duty cycle. It permit a high performance exceeding conventional spike and square wave pulsers. Furthermore several other operations on the received echo, such as amplification, filtering and increasing the signal/noise ratio are also possible.

The transducer's sensitivity versus frequency with air as backing medium and distilled water as propagating medium is given on the figure 16.

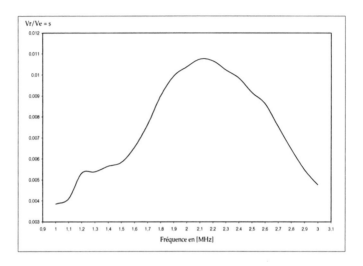

Fig. 16. Transducer's sensitivity versus excitation frequency

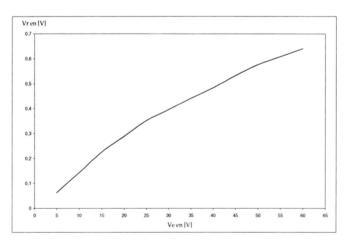

Fig. 17. Transducer's sensitivity versus excitation amplitude

Similarly, variations of transducer's sensitivity versus excitation amplitude are given in figure 17. The slope shows the value of $\xi \approx 0.014$ for the sensitivity.

Variations of the acoustical velocity versus the temperature are shown in figure 18 for distilled water and in figure 19 in the case of paraffin oil. They are close to 1500 *m/s* for distilled water and to 1550 *m/s* for paraffin oil.

In the same way, the acoustical velocity value surrounds 1160 *m/s* in the case of ethanol and 1530 *m/s* in sea water.

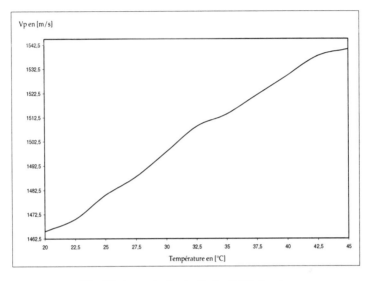

Fig. 18. Acoustical velocity in distilled water

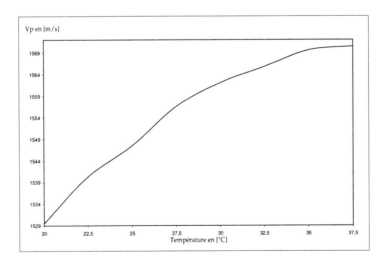

Fig. 19. Acoustical velocity in paraffin oil

6 Conclusion

In this paper, based on previous works a new approach of ultrasonic transducer modelling system is presented. The use of VHDL-AMS language shows the advantage to combine multiphysical domains. The approach can be readily used in current electronic design flow to include distributed physics effects into modelling and simulation process with VHDL-AMS.

The transducer is simulated by the Redwood model, while the medium of propagation is represented by a transmission line which supposes the plane wave theory.

The model allows development of further optimisation with respect to electrical matching and transmitted waveform. It also could be extended to include other phenomena like diffraction and distortion of the acoustic wave propagation in the materials under test.

Usual medium modelling are based on transmission line theory. To perform measurements sensitivity, we can easily adjust in simulation transducer acoustic parameters and also take into account the best parameter for the transducer conception.

The transducer response obtained in simulation shows a good correlation with measurement and indicate that the simulation of an ultrasound sensing device, including both electronics and transducers (electromechanical) is possible using VHDL-AMS.

References

[1] Mason, W.P.: Electromechanical transducers and wave filters, 2nd edn. Van Nostrand, New York (1942)
[2] Kino, G.S.: Acoustic Waves. Prentice-Hall, Englewood Cliffs (1987)
[3] Redwood, M.: Transcient performance of a piezoelectric transducer. J. Acoust. Soc. Amer. 33, 527–536 (1961)
[4] Morris, S.A., Hutchens, C.G.: Implementation of Mason's model on circuit analysis programs. IEEE Trans. Ultrason., Ferroelect., Freq. Contr. 33, 295–298 (1986)
[5] Krimholtz, R., Leedom, D.A., Matthei, G.L.: New equivalent circuits for elementary piezoelectric transducers. Electron. Lett. 6, 398–399 (1970)
[6] Leach Jr., W.M.: Controlled-source analogous circuits and SPICE models for piezoelectric transducers. IEEE Trans. Ultrason., Ferroelect., Freq. Contr. 41, 60–66 (1994)
[7] Püttmer, A., Hauptmann, P., Lucklum, R., Krause, O., Henning, B.: SPICE model for lossy piezoceramic transducers. IEEE Trans. Ultrason., Ferroelect., Freq. Contr. 44, 60–66 (1997)
[8] Maione, E., Tortoli, P., Lypacewicz, G., Nowicki, G., Reid, J.M., Fellow, L.: PSPICE modelling of ultrasound transducers: comparison of software models to experiment. IEEE Ultrason., Ferroelect., Freq. Contr. 46(2), 399–406 (1999)
[9] Hutchens, C.G., Morris, S.A.: A three port model for thickness mode transducers using SPICE II. In: IEEE Ultrasonics Symposium, pp. 897–902 (1984)
[10] Deventer, J.V., Löfqvist, T., Delsing, J.: PSPICE simulation of ultrasonic systems. IEEE Trans. Ultrason., Ferroelect., Freq. Contr. 47, 1014–1024 (2000)

[11] Ghorayeb, S.R., Maione, E., La Magna, V.: Modelling of ultrasonic wave propagation in teeth using PSPICE: a comparison with finite element models. IEEE Ultrason., Ferroelect., Freq. Contr. 48(4), 1124–1131 (2001)
[12] Safari, A., Koray Akdogan, E. (eds.): Piezoelectric and Acoustic Materials for Transducer Applications. Springer, Heidelberg (2008)
[13] Guelaz, R., Kourtiche, D., Hervé, Y., Nadi, M.: Ultrasonic piezoceramic transducer modeling with VDL-AMS IEEE 1076.1. In: Proc. IEEE Sensors, Vienna, Austria (2004)
[14] Guelaz, R., Kourtiche, D., Nadi, M.: A behavioral description with VHDL-AMS of a piezo-ceramic ultrasound transducer based on the Redwood's model. In: Proceedings FDL 2003: Forum on Specification and Design Languages, Frankfurt, German, pp. 32–43 (2003)
[15] Sung, P.H., et al.: The Method for Integrating FBAR with Circuitry on CMOS Chip. In: IEEE International Frequency Control Symposium and Exposition, August 23-27, pp. 562–565 (2004)
[16] Nikoozadeh, A., Wygant, I.O., Lin, D., Oralkan, Ö., Ergun, A.S., Stephens, D.N., Thomenius, K., EDentinger, A.M., Wildes, D., Akopyan, G., Shivkumar, K., Mahajan, A., Sahn, D.J., Khuri-Yakub, B.T.: Forward-looking intracardiac ultrasound imaging using a 1-D CMUT array integrated with custom front-end electronics. IEEE Transactions on Ultrasonics, Ferroelectrics and Frequency Control 55, 2651–2660 (2008)
[17] Branin, F.: Transcient analysis of lossless transmission lines. Proceedings of IEEE 55, 2012–2013 (1967)
[18] IEEE Standard VHDL Analog and Mixed-Signal Extansions, IEEE Std 1076.1-1999, SH94731. IEEE Press, Los Alamitos (1999)
[19] Handbook of Chemistry and Physics, 45th edn. Chemical Rubber Co., Cleveland Ohio
[20] http://www.matec.com/mindt/products/pc_cards/tb-1000/

Application of Laser-Ultrasound to Non-contact Temperature Sensing of Heated Materials

I. Ihara, M. Takahashi, and H. Yamada

Nagaoka University of Technology,
Nagaoka, Japan

Abstract. In this paper we present a new non-contact method with a laser ultrasonic technique for measuring both surface and internal temperature distributions of a heated material. The principle of the temperature measurement is based on temperature dependence of the velocity of the surface acoustic wave (SAW) or longitudinal wave (LW) propagating through a material. An effective method consisting of ultrasonic pulse-echo measurements and an inverse analysis is developed. In the inverse analysis, ultrasonic data are coupled with a one-dimensional finite difference calculation to realize robust and fast estimation of temperature distributions. In addition, a laser-ultrasound system is employed to perform non-contact ultrasonic measurements. To demonstrate the practical feasibility of the method, experiments with an aluminum plate being heated up to 110 °C are carried out. At first, SAWs propagating on the surface of the plate are measured using a laser interferometer based on a photorefractive two-wave mixing. The transit time of the SAW is measured every 0.3 s during heating and used for the inverse analysis to determine the surface temperature distribution of the plate. It is verified that the results determined by the ultrasonic method almost agree with those measured using an infrared radiation camera. Secondly, LWs propagating through inside the heated plate are measured using the laser interferometer. The internal temperature distribution in the plate is then determined from the transit time of the LW. Thus, it has been demonstrated that the proposed method can provide effective non-contact measurements of the transient variation in the temperature distribution of a heated material.

1 Introduction

In the fields of materials science and engineering, there are high demands for monitoring temperature and its distribution in a heated material. This is because such temperature is an important factor that is closely related to the material properties. In particular, it is strongly required to monitor the temperature distribution and its transient variation of the material being processed at high temperatures because the temperature state crucially influences the quality and productivity of final products. Such temperature monitoring is needed for not only the processed material but also the processing machines. For example, in casting or forming processes for metals and polymers, real-time information on temperature gradient inside the die or mould could be useful for realizing an effective process control. Although conventional thermocouple techniques are widely used for temperature measurements, they are not always acceptable for monitoring a transient variation of the internal temperature of the die or mould because of its relatively slow time-response in measurement and its limitation in obtaining the spatial distribution of temperature. It is also important to monitor transient temperature on the surface of a heated material because such a transient variation in the surface temperature gives useful information relating to the internal temperature state of the material. Thertefore, the real-time monitoring of the surface temperature distribution could be beneficial for developing an effective process control for many types of industrial materials productions. Unfortunately, conventional thermocouple techniques are not suitable for precise measurements of the surface temperatures because of atmospheric effects in measurement and difficulty in installing the thermocouples to the material surface. To overcome such problem, an infrared radiation technique is used for surface temperature measurements. Since this method enables noncontact measurements, it is convenient for on-line measurements of surface temperature. In this method, however, accurate temperature measurements are often hindered by the different emissivity and reflection of infrared radiation from other heat sources. Such effects due to the emissivity and reflectivity result in the deterioration of measurement accuracy. In this paper we present an alternative method for non-contact temperature measurements using a laser-ultrasound technique.

2 Motivation

Ultrasound, due to its high sensitivity to temperature, is expected to be an alternative method for temperature measurement. Because there are some advantages in ultrasonic measurements, such as non-invasive and faster time response, several studies on the temperature estimations by ultrasound have been made [1-7]. In our previous works [8, 9], an inverse analysis method based on ultrasonic pulse-echo measurements with longitudinal wave (LW) was developed and applied to the measurements of internal temperature distributions of heated thick plates. It was successfully demonstrated in the works that the internal temperature distribution of the plate can be determined from the transit time of the LW propagating across the thickness of the plate. Although the feasibility of such ultrasonic temperature monitoring of heated materials has shown, most of conventional ultrasonic sensors used for the monitoring are not available for high temperature use. This is basically because piezoelectric materials used for ultrasonic transducers cannot be sustained against high temperatures around the Currie point. Therefore, from a practical point of view, use of a non-contact technique making possible ultrasonic measurements at high temperatures is strongly required. To meet the requirement, we use a laser-ultrasonic technique which provides non-contact ultrasonic measurements. The another advantage of using a laser-ultrasonic technique is to be able to measure surface acoustic waves (SAWs) on a material surface as well as bulk waves such as LWs or shear waves. Since the velocity of SAW propagating on a material surface also exhibits temperature dependence [10, 11], it is highly expected that an ultrasonic inverse method similar to that used in the work [9] should be effective to determine surface temperature distribution, too. It is expected that such temperature measurements technique with SAW can be applied to any material with a shiny or mirror-finished surface to which the infrared radiation technique cannot be applied. In this work, a non-contact method with a laser-ultrasonic technique for monitoring both surface and internal temperature distributions is presented. At first, an effective method consisting of a SAW velocity measurement and an inverse analysis coupled with a one-dimensional finite difference calculation [12] is presented for determining surface temperature distribution quantitatively. To demonstrate the feasibility of the method, surface temperature distributions for aluminum plate are examined. Next, the same laser-ultrasonic technique is applied to the monitoring of internal temperature distribution and its variation in the heated plate. A laser interferometer based on a photorefractive two-wave mixing [13, 14] is used to measure the SAW and LW in the plate during heating.

3 Principle of Temperature Measurement by Ultrasound

It is known that the velocity of an ultrasonic wave propagating through a medium changes with the temperature of the medium. The principle of temperature measurement by ultrasound is based on the temperature dependence of the ultrasonic wave velocity. Assuming a one-dimensional temperature distribution in a medium, the transit time of an ultrasonic wave propagating in the direction of the temperature distribution can be given by

$$t_L = \int_0^L \frac{1}{v(T)} dx, \quad (1)$$

where L is the propagation distance and $v(T)$ is the ultrasonic velocity which is a function of temperature T. In general, the temperature dependence of ultrasonic velocity depends on material properties and may have an approximately linear relationship with temperature for a certain temperature range. When the single side of the medium is heated, the temperature distribution in the medium can be given as a function of location x and time t. Such a temperature distribution $T(x, t)$ is subjected to the thermal boundary condition of the heated medium. On the basis of Eq. (1), if an appropriate inverse analysis with a proper boundary condition is used, it could be possible to determine the temperature distribution from the transit time t_L measured for the heated medium. In fact, the validity of such ultrasonic determination of the temperature distribution was successfully demonstrated through an experiment with a heated silicone rubber plate in our previous work [8]. Recently, an improved ultrasonic inverse method coupled with a one-dimensional finite difference calculation has been developed to determine a one-dimensional temperature distribution inside materials being heated [9]. The advantage of the method is that no information on the thermal boundary condition at the heating surface is needed for inversion. In this work, we have modified the ultrasonic inverse method to be adapted to surface temperature determination using SAW velocity, as shown in the next chapter.

4 Inverse Analysis for Quantitative Determination of Temperature Distribution

Figure 1 shows a schematic of a one-dimensional temperature distribution on the surface of a flat plate whose single side is uniformly heated. To investigate the temperature distribution, we consider a one-dimensional unsteady heat conduction with a constant thermal diffusivity. Assuming that there is no heat source in the plate, the equation of heat conduction in the x-direction on the surface is approximately given by [15]

$$\frac{\partial T}{\partial t} = \alpha \frac{\partial^2 T}{\partial x^2} \quad (2)$$

where T is the temperature, x is the distance from the heating surface A, t is the elapsed time after the heating starts, and α is the thermal diffusivity. The temperature distribution can be estimated by solving Eq. (2) under a certain boundary condition. In actual heating processes, however, the thermal boundary condition at the heating surface A is often unstable and unknown, and therefore, the surface temperature distribution is hardly determined from Eq. (2). To overcome the problem, we have developed an effective method consisting of a SAW measurement

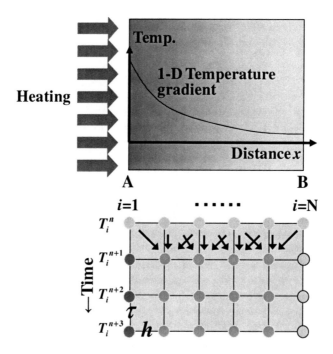

Fig. 1. Schematic of a one-dimensional temperature distribution on the surface of a plate whose single side is being heated (upper), and its finite difference model (downside)

and an inverse analysis coupled with a one-dimensional finite difference calculation. A one-dimensional finite difference model composed of large numbers of small elements and grids is used for analyzing the temperature distribution in the x-direction shown in figure 1. We assume an initial condition that the surface of the plate has a uniform temperature T^n before heating. Using the concept of finite difference calculation, when the single side of the plate starts to heat at the time step n, the temperature of each grid point on the surface at the time step $n+1$, which is a very short time elapsed after the time step n, can be given by [16]

$$T_i^{n+1} = T_i^n + r\left(T_{i+1}^n + T_{i-1}^n - 2T_i^n\right) \quad (i = 2, \cdots, N-1) \tag{3}$$

$$r = \frac{\alpha \tau}{h^2} \tag{4}$$

where N is the total number of grid points and i and n are indices corresponding to the spatial coordinate and consecutive times, respectively. T_i^n is the temperature of each grid point i at the time step n. The coefficient r given by Eq. (4) is known as the von Neumann stability criterion and taken to be less than 0.5 so that we can obtain a stable and proper value for the temperature of each grid point [16, 17], where, τ is the time interval and h is the spatial interval between adjoining grid points. We define $i=1$ as the heating surface A and $i=N$ as the other side B that has no heat source. It should be noted that the temperatures at the two ends, T_1^{n+1} and T_N^{n+1}, are not given in Eq. (3). Since we can calculate the temperatures T_i^{n+1} ($i=2, \cdots, N-1$) from Eq. (3), it is now required to obtain the temperatures at both grid points, T_1^{n+1} and T_N^{n+1}, so that the temperature distribution between A and B on the surface at the time step $n+1$ could be fully determined. It may be reasonable to assume that the temperature T_N^{n+1} can be obtained because such temperature of a low-temperature side can easily be measured using any conventional technique such as a thermocouple or infrared radiation. However, the temperature at the heating surface, T_1^{n+1}, is difficult to measure. Fortunately, it can be possible to estimate T_1^{n+1} if the finite difference calculation is coupled with the transit time of ultrasound propagating through the distance between the two ends, A and B. Using a concept of trapezoidal integration, the transit time t_L given in Eq. (1) can be approximately calculated from

$$t_L = \frac{1}{2} h \left(\frac{1}{v_1^{n+1}} + \frac{1}{v_N^{n+1}} \right) + h \sum_{i=2}^{N-1} \frac{1}{v_i^{n+1}} \tag{5}$$

Using Eq. (5) and the relation between temperature and SAW velocity, the temperature of the heating surface at the time step $n+1$, T_1^{n+1}, can be given by

$$T_1^{n+1} = -\frac{1}{a\left[\dfrac{2t_L}{h} - \left(\dfrac{1}{v_N^{n+1}} + 2\sum_{i=2}^{N-1} \dfrac{1}{v_i^{n+1}}\right)\right]} + \frac{b}{a} \tag{6}$$

where t_L is the transit time measured at the time step $n+1$. It should be noted that Eq. (6) is derived under the assumption that the temperature dependence of SAW velocity has a linear relation shown as

$$v(T) = -aT + b \tag{7}$$

where a and b are constants determined experimentally. The SAW velocity v_i^{n+1} at each grid point in Eq. (6) can be estimated from Eq. (7) since the temperature at each grid point can be obtained from Eqs. (3) and (4). The temperature T_N^{n+1} is considered to be known, as mentioned above. Therefore, the temperature T_1^{n+1} can then be determined from Eq. (6) when the transit time t_L is given as a measured value. Once the temperatures of all grid points between A and B at the time step $n+1$ are determined, we can then determine the temperature distribution at the time step $n+2$ in the same procedure using the transit time t_L measured at the time step $n+2$. Thus, we can continuously obtain the temperature distribution on the surface as long as the SAW measurement is continued and the temperature at B is known.

5 Experimental

A schematic of the experimental setup used is shown in figure 2. This system provides simultaneous measurements of both SAW and LW of a heated plate. An aluminum plate of 30 mm thickness is used as specimens. Two SAWs, SAW A and SAW B, are generated at positions A and B, respectively, using a pulse laser generator (Nd:YAG, λ=1064 nm, energy 200 mJ/pulse, pulse width 5 ns) with a polarizing beam splitter, and detected at position C using a laser interferometer based on a photorefractive two-wave mixing (Nd:YAG, λ=532 nm, energy 200 mW). The advantage of using such a laser interferometer is that it provides robust displacement measurements for a rough surface. It is noted that a LW is also generated at B and the reflected echo from the backside of the plate is detected at

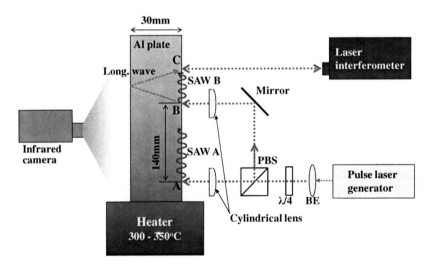

Fig. 2. Schematic of the experimental setup and the laser-ultrasonic system used for measuring both SAWs and LWs of a plate being heated

C. Both SAW and LW measurements are continuously performed, while a bottom side of the plate is heated by contacting with a heater of 300 °C. Figure 3 shows the measured echoes at C on the plate during heating. We can see not only two SAW echoes, SAW A and SAW B, but also two LW echoes, LW 1 and LW 2. It is noted that the two LWs are the 1st and 2nd echoes reflected from the backside. The center frequency of each wave is about 2 MHz. Considering the difference in time delay between SAW A and SAW B, the transit time t_L of the SAW propagating through the distance between A and B can be determined. In this experiment, the propagation distance is precisely estimated to be 139.23 mm from the SAW velocity and transit time in advance. The transit time of the LW propagating through the thickness of the plate is also determined from the difference in time delay between LW 1 and LW 2. To obtain the reference value of the surface temperature distribution of the plate, an infrared camera is used in the backside. The ultrasonic echoes are continuously acquired every 0.3 s with a PC-based real-time acquisition system. Spurious fluctuation due to electrical noise in ultrasonic measurements is reduced by taking the average of ten ultrasonic signals.

Figure 4 shows an infrared radiation image of temperature distribution on the aluminum plate at 180 s after heating started. We can clearly see temperature distribution on the plate surface. A and B in the image correspond to the locations shown in figure 2 and L denotes the area for surface temperature estimation. It is noted that the temperature should be increased with the elapsed time and the one-dimensional temperature distribution along the L and its variation during heating is estimated in this work.

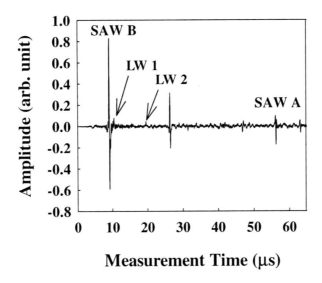

Fig. 3. Measured ultrasonic echoes at position C on the aluminum plate during heating

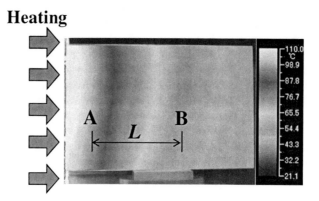

Fig. 4. Temperature distribution on the aluminum plate at 180 s after heating started, measured using an infrared radiation camera

6 Surface Temperature Measurements

Figure 5 shows the changes in the waveforms of the two SAW echoes with the elapsed time after heating started. We can clearly see in Fig. 5 that the waveform of SAW A which is generated at a higher temperature area moves to the right side with an increase in elapsed time, while the waveform of SAW B which is generated at a lower temperature area almost remains as it is. This is because of difference in temperature rise between the location A and B.

Figure 6 shows the variations in the temperatures at A and B and the transit time of the SAW during heating. These temperatures are measured using the infrared camera. As we expected, the temperatures start to increase immediately after the heating starts. The rate of temperature rise at the beginning of heating increases markedly at A where is near the heating surface, while the rate at B where is far from the heating surface increases gradually. The transit time of the SAW is then precisely determined by calculating the cross-correlation between the two echoes, SAW A and SAW B. We can see that the transit time increases markedly with the rise in the temperature of the plate. A small deviation in the measured transit time was observed owing to the distortion in the measured waveforms of the SAWs. To reduce the deviation, the variation in transit time is smoothed by taking the average of adjoining five data. The smoothed curve is then used for the inverse analysis to determine the temperature distribution on the plate surface. Prior to the inverse analysis, the relationship between the SAW velocity and temperature for the aluminum is measured using the laser–ultrasonic system. The result is shown in figure 7 [12]. It is found that the SAW velocity change is almost linear with temperature in the range from 25 to 200 °C, and therefore, the temperature dependence is approximately given by [12]

$$V_{SAW} = -0.7557\, T + 2981.7 \text{ (m/s)} \tag{8}$$

This relation is used for the inverse analysis. In addition, the surface temperature at B measured using the infrared camera is used as a known value for the analysis, as mentioned earlier in this chapter. In the inverse analysis, the following values are used: $\alpha = 96.8 \times 10^{-6}$ m^2/s, $\tau = 0.3$ s, and $h = 10$ mm.

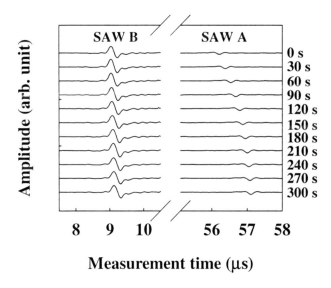

Fig. 5. Variations in the waveforms of the two SAW echoes, ASW A and SAW B, with the elapsed time after heating started

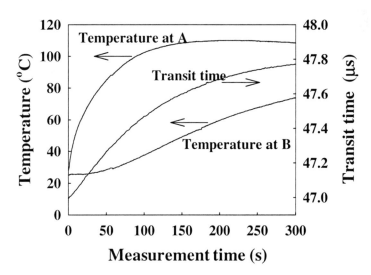

Fig. 6. Variations in the temperatures and transit time of the SAW on the plate during heating

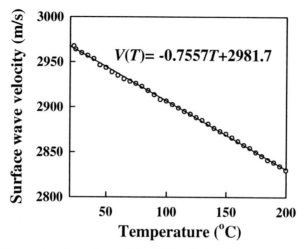

Fig. 7. Temperature dependence of the SAW velocity of the aluminum used [12]

Figure 8 shows the estimated surface temperature distributions for the aluminum plate and their variations with the elapsed time after the heating started, where the results estimated using the ultrasonic technique are compared with those measured using the infrared camera. It can be seen from figure 8 that both temperature distributions determined by ultrasound and infrared radiation almost agree with each other. There are small discrepancies between them in the early stage of heating. Although the reason for the discrepancy is not clear at this moment, it is probable that the results obtained using the infrared camera contain some measurement error due to an uncertainty of the emissivity and reflectivity of infrared radiation.

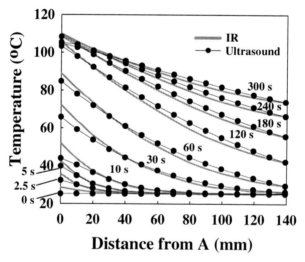

Fig. 8. Surface temperature distributions and their variations with the elapsed time, measured by ultrasound and infrared radiation methods

7 Internal Temperature Measurements

Next, we attempt to measure the internal temperature of the plate. As we mentioned earlier in the chapter of motivation, it was already demonstrated in our previous works [8, 9] that the internal temperature distribution of the heated plate can be successfully measured by the inverse analysis method based on ultrasonic pulse-echo measurements with LW propagating across the thickness of the plate. However, it should be noted that there is some difficulty in applying the ultrasonic technique used in the previous works [8, 9] to the evaluation of materials being heated at higher temperature, because of the limitation of using contact-type ultrasonic transducers which cannot be sustained at such higher temperatures. In this work we have applied the laser ultrasonic technique to the internal temperature measurement so that we could overcome the problem in measurements at higher temperatures. Basically, the LW measurement for the aluminium plate is performed with the experimental setup shown in figure 2 and the inverse analysis method developed in the previous work [9] is employed to determine the internal temperature distribution. It is noted that the inverse method is very similar to that described in the chapter 4 in this paper.

Figure 9 shows the changes in the waveforms of the two LW echoes with the elapsed time after heating started. LW 1 and LW 2 are the 1st and 2nd echoes reflected at the backside of the plate, respectively. It means that they make a round trip and two round trips through the thickness of the plate, respectively. Although the signal-to-noise ratio of the two echoes is not high, we can see that the echoes move to the right side with an increase in elapsed time. This is because of temperature rise of the plate. The transit time of the LW through the plate is

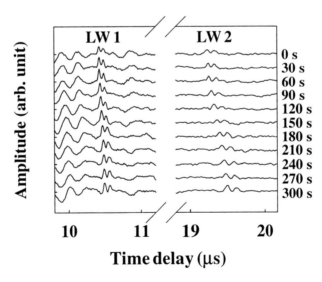

Fig. 9. Variations in the waveforms of the two LW echoes, LW 1 and LW 2, with the elapsed time after heating started

determined from the difference in time delay between the LW 1 and LW 2 by calculating the cross-correlation between the two echoes. Figure 10 shows the variations in the transit time of the LW and the temperature at the backside of the plate measured using the infrared camera. Some deviation in the transit time is observed because of the relatively low signal-to-noise ratio in the waveforms as shown in Fig. 9. The transit time is then used for inverse analysis [9] to determine the internal temperature distribution of the plate. Prior to the inverse analysis, the relationship between the LW velocity and temperature for the aluminum is measured. The result is shown in figure 11. Since the relationship is almost linear in the range from 25 to 100 °C, the temperature dependence of the LW velocity is approximately given by

$$V_{long} = -1.052\ T + 6381.5\ (m/s) \tag{9}$$

This relation is used for the inverse analysis. In this analysis, the surface temperature at the backside measured using the infrared camera is used as a known value, and the following values are used: $\alpha = 96.8 \times 10^{-6}$ m^2/s, $\tau = 0.3$ s, and $h = 10$ mm.

Figure 12 shows the estimated internal temperature distribution and its variations with the elapsed time after the heating started. It should be noted here that the temperature distribution is estimated in the area across the thickness of the plate at the middle region between B and C in figure 2, where the distance between B and C is 24 mm. It can be seen from Fig. 12 that there is almost no temperature gradient inside the plate and the temperature increases gradually as the heating time proceeds. Because the accuracy in the estimated temperature

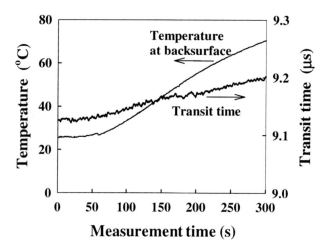

Fig. 10. Variations in the temperature and transit time of the LW through the thickness of the plate during heating

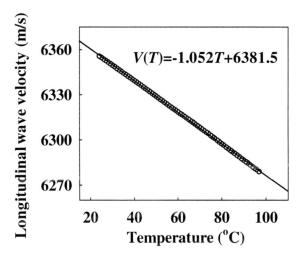

Fig. 11. Temperature dependence of the LW velocity of the aluminum used

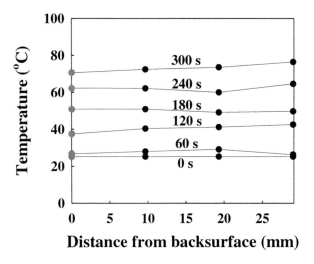

Fig. 12. Internal temperature distribution through the thickness of the plate and its variation with the elapsed time, estimated by ultrasound

gradient may be not good enough due to the low signal-to-noise ratio in the waveforms shown in figure 5, the validity of the result is now being examined. However, it is considered that the tendency in the temperature distribution seems reasonable from the empirical knowledge. Thus, it is demonstrated through the experiments with the heated aluminum plate that the ultrasonic method using a laser-ultrasound technique provides non-contact monitoring of both internal and surface temperature distributions of a material being heated.

8 Conclusion

A new non-contact method with a laser-ultrasonic technique for measuring a one-dimensional temperature distribution of a heated material is presented. The advantage of this method is that the use of a laser-ultrasonic technique enables non-contact measurements of both internal and surface temperature distributions. The practicability of the proposed method is demonstrated through experiments with aluminum plate heated up to 110 °C. Although further study is necessary to improve the robustness and accuracy in measurement, it is highly expected that the method will be used for the on-line monitoring of the transient temperature variation of the material being processed at high temperatures. In particular, the advantages of using ultrasound, such as a wide-use, fast time-response, less expensive, and safety, are quite attractive to industrial applications. More experiments are being made to demonstrate the usefulness of the present method at higher temperatures up to 1000 °C.

References

[1] Degertekin, F.L., Pei, J., Khuri-Yakub, B.T., Saraswat, K.C.: In-situ Acoustic Temperature Tomography of Semiconductor Wafers. Appl. Phys. Lett. 64, 1338–1340 (1994)

[2] Simon, C., VanBaren, P., Ebbini, E.: Two-Dimensional Temperature Estimation Using Diagnostic Ultrasound. IEEE Trans. on Ultrasonics, Ferroelectrics, and Frequency Control 45(4), 1088–1099 (1998)

[3] Chen, T.-F., Nguyen, K.-T., Wen, S.-S., Jen, C.-K.: Temperature measurement of polymer extrusion by ultrasonic techniques. Meas. Sci. Technol. 10, 139–145 (1999)

[4] Balasubramainiam, K., Shah, V.V., Costley, R.D., Boudreaux, G., Singh, J.P.: High temperature ultrasonic sensor for the simultaneous measurement of viscosity and temperature of melts. Review of scientific Instruments 70(12), 4618–4623 (1999)

[5] Huang, K.N., Huang, C.F., Li, Y.C., Young, M.S.: High precision, fast ultrasonic thermometer based on measurement of the speed of sound in air. Review of scientific Instruments 73(11), 4022–4027 (2002)

[6] Mizutani, K., Kawabe, S., Saito, I., Masuyama, H.: Measurement of temperature distribution using acoustic reflector. Jpn. J. Appl. Phys. 45, 4516–4520 (2006)

[7] Tsai, W.-Y., Chen, H.-C., Liao, T.-L.: An ultrasonic air temperature measurement system with self-correction function for humidity. Meas. Sci. Technol. 16, 548–555 (2005)

[8] Takahashi, M., Ihara, I.: Ultrasonic Monitoring of Internal Temperature Distribution in a Heated Material. Jpn. J. Appl. Phys. 47, 3894–3898 (2008)

[9] Takahashi, M., Ihara, I.: Ultrasonic Determination of Temperature Distribution in Thick Plates during Single Sided Heating. Mod. Phys. Lett. B 22, 971–976 (2008)

[10] Nishimura, K., Shigekawa, N., Yokoyama, H., Hohkawa, K.: Temperature dependence of surface acoustic wave characteristics of GaN layers on sapphire substrates. Jpn. J. Appl. Phys. 44, 564–565 (2005)

[11] Wang, S., Harada, J., Uda, S.: A wireless SAW temperature sensor using langasite as substrate material for high temperature applications. Jpn. J. Appl. Phys. 42, 6124–6127 (2003)

[12] Takahashi, M., Ihara, I.: Quantitative Evaluation of One-Dimensional Temperature Distribution on Material Surface Using Surface Acoustic Wave. Jpn. J. Appl. Phys. 48 (2008) (in press)

[13] Pouet, B.F., Ing, R.K., Krishnaswamy, S., Royer, D.: Heterodyne interferometer with two-wave mixing in photorefractive crystals for ultrasound detection on rough surfaces. Appl. Phys. Lett. 69, 3782–3784 (1996)

[14] Pouet, B.F., Breugnot, S., Clémenceau, P.: Robust Laser-Ultrasonic Interferometer Based on Random Quadrature Demodulation. In: AIP Conf. Proc., vol. 820, pp. 233–239 (2006)

[15] Myers, G.E.: Analytical Methods in Conduction Heat Transfer, p. 200. McGraw-Hill, New York (1971)

[16] Buchanan, J.L., Turner, P.R.: Numerical Methods and Analysis, p. 422. McGraw-Hill, New York (1992)

[17] Press, W., Teukolsky, S., Vetterling, W., Flannery, B.: Numerical Recipes in C++, p. 847. Cambridge University Press, Cambridge (2003)

Vision Sensor with an Active Digital Fovea

Donald G. Bailey[1] and Christos-Savvas Bouganis[2]

[1] School of Engineering and Advanced Technology
  Massey University, Palmerston North, New Zealand
[2] Department of Electrical and Electronic Engineering
  Imperial College London, London, United Kingdom

Abstract. Foveated image sensors have a variable spatial resolution enabling a significant reduction in image size and data volume. In this work, the requirements for a foveated sensor within an active vision system are analysed. Based on these requirements, the constraints on the resolution mapping function are determined, and a range of Cartesian based mapping schemes investigated. The results demonstrate that separable mappings, which independently map two orthogonal dimensions are efficient to implement using an FPGA. The computational requirements for an L_2 or Euclidean radial mapping are significantly higher, although this yields more natural looking low-resolution images. A compromise is based on using an L_∞ radial mapping. The requirement to process data as it is streamed from the camera necessitates implementing the warping using a forward mapping rather than the more common reverse mapping. The implications of this for FPGA implementation are described.

1 Introduction

In recent years image sensors have reached very high capacity making 5 to 10 megapixels in one sensor a commodity. This allows the acquisition of high-resolution images, which usually has a positive impact on the overall performance of many computer vision algorithms. However, it also creates a computational burden in processing systems when real-time constraints have to be met and the whole information from the sensor has to be processed. Moreover, in certain applications, such as tracking and pattern recognition, it is not so important to maintain the same resolution across the image sensor as to have a wide field of view and have high resolution only on specific regions of the sensor. Thus, a trade-off between high resolution and processing power has been created.

While it is possible to process such data in real time using dedicated hardware or a DSP microprocessor, any algorithm that requires multiple frames will require significant off-chip memory access, with its consequent bandwidth bottleneck. To enable on-chip storage, the volume of data, hence image size, must be reduced considerably. This chapter expands on earlier work by the authors (Bailey and Bouganis 2008).

1.1 Foveal Vision

The simplest way to reduce the data volume is to reduce the image size, and hence resolution. The consequence of this is a loss of information that may be critical in many applications. Often, high resolution is only critical in small regions of the image. A foveated window, inspired by the human visual system, provides a balance between high resolution, and large data volume. Within the window, the high resolution is maintained in the centre, with the resolution decreasing towards the periphery. Compared to a uniform resolution image, the increase in resolution in the centre comes at the expense of a decrease in resolution at the periphery.

Multi-resolution techniques are usually employed to address the above problem, where recently space-variant or foveating image sensors have been introduced that address the problem in its origin. These sensor architectures have variable spatial resolution across the surface of the sensor targeting data reduction without a severe impact on the final performance of the application. In (Martinez and Altamirano 2006) Martinez and Altamirano have demonstrated that a data reduction by a factor of 22 can be achieved without significant degradation in the performance of their tracking algorithm. Similarly, (Bailey and Bouganis 2009) show that tracking performance is not significantly affected after reducing the data volume by a factor of 64. An important part of such systems is the use of active vision techniques to ensure that the high resolution part of the sensor corresponds to the region of the scene where it can be most effective.

1.2 Potential Applications

Pattern recognition and tracking are two of the domains where a foveating image sensor has been successfully employed. Wilson and Hodgson (Wilson and Hodgson 1992a; b) addressed the problem of pattern recognition using a space-variant sensor. They exploited the rotation and scale invariance properties of a log-polar mapping combined with the reduction in data volume to efficiently represent patterns as templates, and to recognize images through a relatively simple template matching process for quite complex patterns. This approach requires an accurate and dynamic positioning of the fovea, for which the centre of gravity of the input Cartesian image was used. Scale invariance may be achieved with other mappings by dynamically matching the resolution (and mapping function) to the size of the object being recognized.

A space-variant sensor system was employed in (Cui et al. 1998; Xue and Morrell 2002) for tracking. Tracking requires positioning the fovea region on the object of interest. As the object moves from the centre of the fovea, this may be detected, and the position of the fovea dynamically adjusted in the next frame to enable the object to be tracked. By predicting the motion, a foveated sensor is able to maintain the target within the high-resolution region, and effectively track at this high resolution (Bailey and Bouganis 2009) while efficiently processing much smaller images.

A further application is in image compression (Wang and Bovik 2001). A foveated image is able to maintain resolution where detail is important, while reducing the resolution, or increasing the compression, in regions where detail is not required.

For many of these applications, a saliency detection algorithm (Itti et al. 1998; Liu et al. 2006) is required to identify regions of interest within the image. These algorithms model the human attention model and estimate the regions of the image that attract the human attention and thus should be given more resolution than the other regions. These algorithms are typically used on the original high-resolution images.

Different applications may require a variety of mapping functions to be incorporated in the system so that the mapping function may be dynamically selected at run-time. It is also necessary to adapt the range of standard image processing techniques to operate on the low-resolution images.

1.3 Requirements Analysis

Many methods have been proposed for acquiring variable resolution images. In this section, a set of key requirements of a low power active foveal vision system is proposed and discussed. FPGAs are becoming increasingly used for low power embedded vision systems because they are able to exploit parallelism to enable a lower clock frequency to be used than a serial processor. For this reason, this chapter focuses primarily on an FPGA based implementation. The requirements are:

1) The system must be able to dynamically control the position of the fovea. In most applications, the vision system has to handle dynamic scenes. Thus, the regions of the image that require high resolution are not known in advance, leading to the requirement of repositioning of the fovea from one frame to the next. This may be accomplished either by mechanically scanning the fovea, or by electronically repositioning the fovea region over the region of interest.

2) The system must be able to achieve a significant reduction in the volume of data. Even though modern systems can readily access large off-chip memories, the communication between the memory and the computational unit is often the bottleneck for real-time performance. A significant reduction in the volume of the data will reduce the transfer times, and will even allow multiple images to be stored on-chip. For example, a 64×64 output image requires only 4 kbytes. Therefore, in an FPGA implementation, a monochrome image could be held in a single Virtex 5 Block RAM (Xilinx 2008), and a colour image could be held in a single Stratix IV M144K block (Altera 2008).

3) The system has to be able to create the foveated image as fast as possible for high frame rate processing to be achieved. This sets an upper limit to the computational complexity of the system that performs the conversion to a foveated image. Ideally, to minimize the latency the image should be converted as the data is streamed from the sensor onto the FPGA.

4) The system must be able to successfully perform the image processing operations on the low-resolution images, despite the large targeted reduction in the volume of the data. Successful means that existing algorithms must be able to be adapted (or new algorithms devised) that give similar results to processing the original high-resolution image but with significant reduction in processing time. In general, what is considered successful will be application dependent, however, any small reduction in the quality of the results must come with the benefit of significant reductions in processing time.

5) If multiple foveal mappings are available, there must be some way of dynamically selecting the best mapping for application. This may be as simple as tuning the mapping function based on the size of the object of interest, to dynamically optimizing the mapping function as part of the active vision process.

2 Related Work

Many configuration topologies have been introduced in the literature for spatially variant imaging. Most are inspired by the human vision, having a resolution that decreases with the distance from the centre of the sensor.

2.1 Methods of Acquiring Variable Resolution Images

2.1.1 Optical Approaches

In (Kuniyoshi et al. 1995), Kuniyoshi et al. achieved a variable spatial resolution by using a regular CCD sensor combined with a specifically designed lens that mimics the acuity of the human visual system. The authors focused on achieving a projection curve that would mimic the human eye projection characteristics and at the same time the current vision applications could be applied to the acquired image data. The authors consider three types of projection curves. In the fovea region, the authors employed a standard projection, since this region would be mainly used to extract local image features. For the periphery curve, a spherical projection was selected, since this area focuses on motion analysis and detection and for the extraction of global features from the environment in general. The chosen projection type for the periphery is supported by the fact that it simplifies the optical flow analysis for separating the target motion from ego-motion and also can help with navigation. In order to connect the two distinct areas, the authors propose the use of a log-polar projection. They concluded that the design of the optical system with the above characteristics was challenging and some digital processing was a necessity.

More recently, Hua and Liu (Hua and Liu 2008) proposed an approach that uses two off-the-shelf image sensors combined with a beam splitter to create a foveating imaging system. In this work, the authors' aim was to utilize the spatially variant resolution characteristic of the human visual system along with the space variant characteristics of contrast and color sensitivity in order to reduce the data bandwidth and maximize the information throughput. An imaging system based on a dual-sensor approach was presented which employed two detectors that shared the same entrance pupil. Moreover, the two detectors were coupled using a two-axis MEMS scanner to enable the high-resolution sensor to be mapped anywhere within the field of view of the low-resolution sensor. The authors reported a 90.4% bandwidth reduction, compared to a uniform sampling system.

2.1.2 VLSI Approaches

Targeting a small form factor and reduced power consumption, researchers have focused on manufacturing image sensors that have variable spatial resolutions. In (Etienne-Cummings 2000) the authors present a 2D foveated silicon retina which is realized in a single chip. The presented system exhibits two static areas with different resolutions. The fovea is in the center of the sensor and contains a 9×9 dense array, where the surrounding area implements the peripheral region and contains a sparse array of 19×17 larger photoreceptors. Moreover, the proposed system implements some functions of the primate retina such as direction-of-motion detection in the retina and the localization of spatiotemporal edges in the periphery.

The main drawback of the VLSI approach is the fixed topology. In order for the system to perform tracking, a pair of motors is required. This problem is overcome by Vogelstein et al. (Vogelstein et al. 2004) by dynamically combining adjacent sensing elements to achieve varying spatial resolution. Their 80×60 image sensor has an address-event representation communication protocol which is interfaced with a digital microprocessor and up to four aVLSI chips that implement the integrate-and-fire neurons. By suitable configuration of the aVLSI chips, a fovea region can be realized across the image sensor, and by pooling events, a lower resolution may be achieved in the periphery.

2.1.3 Software Emulation

Emulating a variable resolution sensor using software has been adopted by many researchers due to the low cost and high flexibility that it offers. However, for real-time applications, the required processing time limits its applicability.

2.1.4 Hardware Emulation

A widely used approach is the use of a standard uniform resolution CCD sensor combined with either a VLSI chip or FPGA that maps the image to a topology that emulates a variable resolution sensor. Camacho et al. (Camacho et al. 1998) interface a progressive scan CCD camera with an FPGA to perform an adaptive mapping of the fovea. The FPGA is responsible for processing the data and constructing the multi-fovea image. It should be noted that the authors propose the use of many types of fovea that exhibit different spatial resolution capabilities under a log-Cartesian topology. Moreover, all the fovea regions exhibit a rectangular shape.

In (Ovod et al. 2005), the VASI (variable acuity super-pixel imager) system is described. It uses a CCD sensor and an FPGA for real-time image processing. The system supports multiple fovea, but only two levels of resolution are possible.

Arribas and Macia (Arribas and Maciá 1999) proposed an implementation of log-polar mapping using an FPGA that was able to achieve real-time performance. More recently, Martinez and Altamirano (Martinez and Altamirano 2006) proposed an FPGA pipelined architecture that transforms Cartesian images to a foveated image. The authors approximate a log-polar mapping by multiple square regions of different resolutions. They argue that the non-linearities that introduced in the image representation when a log-polar transformation is applied makes current computer vision algorithms such as correlation-based detection or recognition hard to apply.

2.2 Types of Mapping

The various types of foveal mappings that have been proposed in the literature can be categorized into four main types.

2.2.1 Dual Resolution

This type of mapping consists of two Cartesian regions, usually rectangular, of different spatial resolution. This may be achieved by employing two different types of lens (Hua and Liu, 2008) or by explicitly constructing the image sensor to have two sampling regions of different densities (Etienne-Cummings 2000).

2.2.2 Log-Polar Mapping

The log-polar mapping has been one of the most widely used foveated mapping due to its close characteristics to the human retina. In (Wilson and Hodgson 1992a; b) Wilson and Hodgson use a log-polar mapping inspired by the sampling structure of the human retina, for a range of pattern recognition tasks. The main advantage of the log-polar mapping is that the logarithmic radial resolution dependence and polar coordinate systems enable scale and rotation invariance through translation within the low-resolution image if the fovea has been positioned consistently. This invariance can be exploited to simplify many pattern recognition tasks. Traver and Pla (Traver and Pla 2003), consider the trade-offs in the sensor topology when a log-polar mapping is targeted.

Although the log-polar mapping is a widely used mapping, one of its drawbacks is the non-linear effects that the polar coordinate system introduces to the regions of the sensor that do not align with the optical axis. Another problem is the singularity and consequent blind-spot in the centre of the fovea.

2.2.3 Ring Structures

A compromise is to base the mapping on a ring structure like the log-polar mapping, without using polar coordinates. Bandera and Scott (Bandera and Scott 1989), investigate the use of various imager topologies for a foveal vision system with respect to the data structure size and computations. In their work, they propose the use of rectangular and hexagonal lattices with varying resolutions, where the effective resolution of the system decreases with the distance from the center.

2.2.4 Others

Recently, research has focused on sensors that can provide many fovea regions rather than having a fixed topology. Camacho et. al. (Camacho et al. 1996) proposed a multi-resolution topology where the fovea region can be shifted across the sensor emulating the saccadic eye movements. In (Camacho et al. 1998) they generalised this idea to multiple foveal regions and moving object detection.

3 System Architecture

The approach taken in this chapter utilizes a continuously variable spatial resolution from the fovea to the periphery. It uses a standard uniform resolution image sensor and emulates a variable resolution sensor by mapping the uniformly sampled input image to a low resolution foveated image using an FPGA.

To meet the requirements listed in section 1.3, the architecture shown in Fig. 1 is proposed. It makes use of the high-resolution (3 to 10 megapixel) CMOS sensors that are now readily available. The high pixel count enables a wide-angle lens to be used without a loss of resolution compared to that of a standard camera. However, the high pixel count also has the disadvantage of limiting the frame rate.

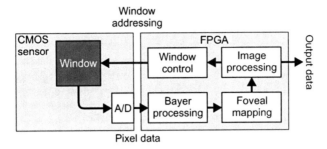

Fig. 1. System architecture

This may be overcome with CMOS sensors, because such sensors allow any arbitrary rectangular region or window of the sensor to be read out. Since the readout window is programmable, it may be positioned anywhere within the field of view of the camera.

Such a sensor enables variable spatial resolution in at least two different ways. The first is the dual resolution approach, which can be provided with a single high-resolution sensor. First the full image may be read out, using binning within the readout circuitry to combine groups of adjacent rows and columns of pixels. This reduces the resolution of the full image, reducing its size enabling it to be read out quickly. This provides the low-resolution "background" image for the dual resolution system. Then a full resolution image may be read out using the window, providing the high-resolution "detail" image. This approach effectively multiplexes the single sensor between it use for the high resolution and low resolution required by a dual resolution system.

The second approach, which will be explored in this chapter, is to read the high resolution data from the readout window of the sensor, and resample the uniform resolution image to provide a variable resolution foveal image for processing.

Digital CMOS sensors have integrated analogue to digital conversion, enabling direct connection to an FPGA. Single chip colour sensors use a Bayer pattern (Bayer 1976) or similar array of coloured filters over each pixel. Each pixel therefore only provides a single colour channel, so adjacent pixels must be interpolated to obtain a full colour image. The image read from the camera is then resampled using the foveal mapping to give an image that is significantly reduced in size. The small foveal image is then processed to give the required output from the complete camera. This processing will depend significantly on the application. The foveal images are also processed to determine the next location for the fovea, which is used to control the position of the readout window in the sensor, closing the feedback loop shown in Fig. 1.

3.1 Fovea Positioning

The ability to read out pixel data from within a predefined region of the sensor is essential for high-speed foveal vision. It enables the position of the high-resolution region to be positioned from one frame to the next under program control.

Without such an ability (for example using an optical approach, or defining the variable sensor geometry at the silicon level) it is necessary to mechanically pan and tilt the camera to position the fovea. Repositioning the window within a frame provides the equivalent of a very fast electronic pan and tilt. As the camera does not move, both the latency and motion blur associated with physically panning or tilting the camera are avoided. The limited angle of view, even of a wide-angle lens, means that some applications may still require a mechanical pan-tilt head. However, in applications where only a limited field of view is required, the need for a pan-tilt head may be eliminated.

3.2 Bayer Processing

There is a wide range of algorithms for interpolating the missing colour values in an image captured using a colour filter array. The demosaicing algorithms are effectively filters, which require data from adjacent rows to provide an output value. Data from previous rows must be cached using row buffers. From an FPGA processing perspective, it is desirable to minimize the required hardware resources.

Nearest neighbour and bilinear interpolation are relatively low cost, requiring one and two row buffers respectively. However, the problem with such simple interpolation is that it is prone to artifacts, particularly around edges and in regions of fine detail.

Many of the more complex algorithms are designed to reduce the artifacts introduced by the simpler methods. Unfortunately, while such algorithms can give good results, both the computational complexity, and the number of row buffers required can be quite large.

One compromise described in (Hsia 2004) uses a small window requiring only two row buffers. The improvement in quality is obtained by detecting edge directionality and weighting accordingly, and using a local gain to account for fine details. In addition to the row buffers, the technique requires two small dividers and three small multipliers. Both the multiplication and division may be implemented efficiently using logic gates on the FPGA (Bailey 2006).

3.3 Foveal Mappings

The main idea of the foveal mapping is to deliberately introduce distortion, so that part of the image (usually the centre) has a high resolution, while the periphery has a lower resolution. There are two alternatives in terms of the distortion mapping function.

One is to base the output on a polar coordinate system, such as that used by the log-polar mapping (Wilson and Hodgson 1992b; Traver and Pla 2003). The foveal mapping in this case performs a rectangular to polar conversion. In such a mapping, the angular resolution is constant, which naturally gives higher spatial resolution at the origin and has radially descreasing spatial resolution. The radial mapping function is designed to maintain a uniform radial to angular aspect ratio.

The alternative is to maintain Cartesian coordinates in both the input and output images. The foveal mapping in this case introduces a distortion, a little like an exaggerated lens distortion or fisheye lens distortion. These distortions result from a non-uniform magnification in the lens. Inspired by this, the input image can be deliberately warped to both maintain the field of view by using a lower magnification in the periphery, while keeping the resolution in the fovea.

Let u be the distance from the centre of the input window and f be the distance from the centre of the foveated image. The foveal mapping can then be defined either in terms of the forward mapping, which gives the position of a pixel in the foveated image in terms of the undistorted input

$$f = \text{map}_f(u), \quad (1)$$

or the reverse mapping which gives the location in the input image in terms of the output

$$u = \text{map}_r(f). \quad (2)$$

The magnification, M, at any point is defined as the ratio between the output and input pixel distances. It may be expressed either in terms of the input pixels,

$$M_f(u) = \frac{f}{u} = \frac{\text{map}_f(u)}{u}, \quad (3)$$

or the output pixels

$$M_r(f) = \frac{f}{u} = \frac{f}{\text{map}_r(f)}. \quad (4)$$

The magnification relates the overall sizes of the input and foveated images, and therefore gives the average size of each output pixel in terms of input pixels. However, since the size of the output pixels is variable, a more useful relationship is the acuity, which is the effective resolution of an output pixel in terms of the uniform input resolution. The acuity, A, is therefore given from the slope of the mapping.

$$A(f) = \frac{df}{du} = 1 \Big/ \frac{d\,\text{map}_r(f)}{df}. \quad (5)$$

When expressed in terms of input pixels, it gives the size of the specified input point in the output image:

$$A(u) = \frac{df}{du} = \frac{d\,\text{map}_f(u)}{du}. \quad (6)$$

There are relatively few constraints on the mapping function. To ensure that an input point maps to only a single point in the foveated image, the mapping must be monotonic. There is little reason for having a higher resolution in the fovea than is in the original input image. (An exception perhaps is to introduce magnification into a dynamic mapping to maintain scale invariance in an object recognition application.) These imply

$$0 < A \le 1 \quad (7)$$

To map an *N*×*N* input window to a *W*×*W* output image also requires that both the centre and edges of the input map to the centre and edges of the output respectively:

$$\text{map}_f(0) = \text{map}_r(0) = 0 \tag{8}$$

And

$$\begin{aligned}\text{map}_f(N/2) &= W/2 \\ \text{map}_r(W/2) &= N/2\end{aligned} \tag{9}$$

If the fovea is in the centre of the image, and the resolution of the fovea corresponds to the input resolution then

$$A(0) = 1.0. \tag{10}$$

Note, this is not actually a constraint; the resolution in the fovea does not have to correspond to the input resolution, and the best resolution does not have to be in the centre of the output image.

Consider a typical input image resolution (sensor readout window size) of 512×512. To give a significant level of data reduction (as defined in section 1.3), a suitable output image size would be 64×64. Three example mappings that match the constraints of Eqs. (7) to (10) are

$$\text{map}_r(f) = f + \tfrac{7}{32} f^2, \tag{11}$$

$$\text{map}_r(f) = f + \tfrac{7}{1024} f^3, \tag{12}$$

and

$$\text{map}_r(f) = f + \tfrac{7}{32768} f^4. \tag{13}$$

The corresponding acuity functions, expressed in terms of the output pixels are shown in Fig. 2.

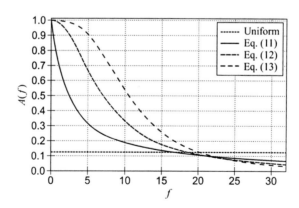

Fig. 2. Acuity for three different fovea mappings compared with uniform downsampling

A couple of observations may be made regarding the mappings. There is a gradation in the width of the fovea from Eq. (11) through to Eq. (13). The resolution of Eq. (11) drops very quickly away from the centre, whereas the mapping of Eq. (13) has several high resolution pixels in the fovea. There is a tradeoff with the size of the fovea, because a larger fovea means the periphery must be fitted into fewer pixels, resulting in a significantly lower resolution at the edges of the image. For example a pixel on the edge of the output image corresponds to an 8×8 block of input pixels for uniform magnification, whereas it corresponds to 15×15, 22×22 and 29×29 blocks respectively for Eqs. (11) to (13).

Another consideration is how to define the distances u and f. Different definitions will affect the nature of the distortion introduced. Let the coordinates relative to the centre of the input window or output image respectively be defined as (x_u, y_u) and (x_f, y_f). A radial Euclidean mapping, using the L_2 distance metric would have

$$u = \sqrt{x_u^2 + y_u^2} \tag{14}$$

And

$$f = \sqrt{x_f^2 + y_f^2} = \text{map}_f(u) = uM_f(u) \tag{15}$$

With a radial mapping, the angle of the point relative to the centre of the image is unchanged. This means that the x and y coordinates are both scaled by the magnification for the given radius:

$$x_f = x_u M_f(u)$$
$$y_f = y_u M_f(u) \tag{16}$$

or equivalently, using the reverse mapping

$$x_u = x_f M_r(u)$$
$$y_u = y_f M_r(u) \tag{17}$$

A computationally simpler transform may be obtained by using the L_∞ or chessboard distance metric,

$$u = \max(|x_u|, |y_u|), \tag{18}$$

again with radial scaling using Eq. (16) or (17).

An even simpler transform may be obtained by considering the mapping separable (rather than radial) and independently mapping x and y:

$$x_f = \text{map}_f(x_u)$$
$$y_f = \text{map}_f(y_u) \tag{19}$$

Note that with the separable mapping there are different magnifications in each of the coordinate directions.

Fig. 3 compares the effects of these three mappings using the mapping function of Eq. (11). The size and shape of the pixels depends on both the mapping, and their position within the output image. The pixels are only square in the centre of

Fig. 3. Comparison of the different mapping methods: the small images are the foveated images, and the large images have remapped them back onto the original to show the acuity. (**a**) original Lena (512×512); (**b**) radial L_2 mapping; (**c**) radial L_∞ mapping; (**d**) separable mapping.

the fovea, and along the diagonals of the separable mapping. The separable transform does have rectangular pixels because of the separability, but the radial mappings result in pixels which are in general diagonal. For the radial mappings, the longest dimension of the pixel (which is radial) is given by the acuity, but the shortest dimension (tangential) is given by the magnification. This means that the pixels in the foveal image generally have better resolution tangentially than

Fig. 4. Comparison of the different mappings using the L_∞ mapping method: the small images are the foveated images, and the large images have remapped them back onto the original. (**a**) uniform downsampling to (64×64); (**b**) mapping of Eq. (11); (**c**) mapping of Eq. (12); (**d**) mapping of Eq. (13).

radially. The change of aspect ratio with position for the radial mappings is more uniform than for the separable mapping, so that standard image processing algorithms are more likely to work correctly on these foveal images. This is less so for the separable mapping.

In the foveal images, objects appear least distorted with the L_2 mapping although horizontal and vertical lines become curved. However, the overall resolution with this mapping is lower because of the regions in the corners without data. The other two mappings show distinct distortion along the diagonals, especially with the separable mapping where straight lines have a distinct corner along the diagonals of the foveal image.

It is also instructive to compare the different size foveas. Fig. 4 compares the mappings of Eqs. (11) to (13) with uniform downsampling. With the larger foveas, the periphery becomes noticeably more compressed, and the consequent increase in aspect ratio at the periphery makes the pixels appear to radiate from the centre in the reconstruction, especially along the diagonals. The pinching distortion along the diagonals also becomes more noticeable.

4 FPGA Implementation

Modern FPGAs have sufficient resources to enable whole applications to be implemented on a single FPGA, making them an ideal choice for embedded real-time vision systems. This section will discuss the issues associated with implementing the foveal mapping on an FPGA.

4.1 Reverse Mapping

Most commonly, the reverse mapping is used to perform image warping. The reverse mapping, defined by Eq. (2), determines the corresponding location in the input image for each output pixel, using some form of interpolation to handle fractions of pixels (Wolberg 1990). This is an advantage when the output pixels must be produced in a particular order, for example when streaming the output for a display. However, there are several problems with using a reverse mapping for implementing the foveal mapping.

First, producing the reverse mapping to implement a warping requires random access to the input image to select the corresponding input pixels for each output pixel. Implementing the mapping in this way would require significant memory for frame buffering, which was contrary to the original design goals.

Storing the image in an intermediate frame buffer also introduces latency into the transformation. With active vision, the outputs are used to control the capture, including the positioning of the fovea. Additional delays will reduce the ability to control the system.

A further problem is that the foveal mapping results in a reduction in resolution. Simply sampling the input image can result in significant aliasing, particularly in the periphery. This must be overcome by prefiltering with an anti-aliasing filter. With the spatially variant resolution, this requires a spatially variant filter, with little or no filtering required in the fovea, and a large window in the periphery.

4.2 Forward Mapping

These problems may be overcome by using a forward mapping, defined by Eq. (1), which determines where in the output image each input pixel maps to. A forward mapping processes the input pixels as they are streamed from the camera. This minimizes the need for frame buffering giving the maximum benefit from the reduction in data volume from the foveal mapping. Since each pixel is processed as it arrives, it can minimize the latency, at least of the mapping stage.

A common problem with using the forward mapping is holes in the output, where there are output pixels with no corresponding inputs (Wolberg 1990). This is not an issue with a foveal mapping because the acuity is always less than or equal to one, resulting in a many to one mapping.

As outlined earlier, associating a single input pixel with each output pixel can result in aliasing. Simulating a low resolution sensor, with larger pixels on the image sensor, requires averaging all of the input pixels associated each output pixel. This averaging is effectively performing the spatially variant filtering required to reduce aliasing.

Such averaging requires maintaining two accumulators for each pixel within the output image. Each pixel is mapped to an output accumulator as determined by the mapping function. If the input pixel spans the boundary of multiple output pixels, then that pixel must be split among the associated output accumulators. The second accumulator counts the number of input pixels accumulated for each output pixel. Dividing one by the other will enable the corresponding output pixel value to be calculated.

4.3 Separable Mapping Implementation

The simplest mapping to implement is the separable mapping. It enables the X and Y directions to be mapped separately, which reduces the logic. This implementation is shown in Fig. 5. The left part of the figure shows the mapping in the X direction, with the circuit repeated on the right for the Y direction.

The mapping is effectively a two-pass algorithm similar to that defined by Catmull and Smith (Catmull and Smith 1980). The first mapping, or pass, assigns the input pixels into the correct output column, and the Y mapping (the second pass) then operates on the column to place each pixel in the correct row. To maintain a low latency, the Y pass is performed on all of the columns in parallel, to

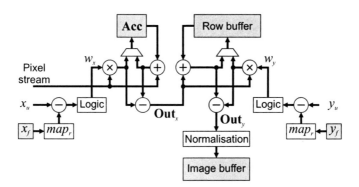

Fig. 5. Separable mapping implementation

enable the pixels to be processed in the order that they are streamed out from the first pass.

Consider first the X mapping. Rather than calculate the mapping algebraically (for example by directly using one of Eqs. (11) to (13)), it is more convenient to store the mapping directly in a lookup table. The simplest would be to use the forward mapping of Eq. (1) to indicate which pixels the input address maps to. However, since there are significantly fewer output pixels than input pixels, the reverse map is considerably smaller. The map_r block of Fig. 5 stores the boundaries in the input image associated with each output position. The incoming address is compared with the mapped address to determine whether or not the incoming pixel straddles two output pixels. If the incoming pixel is completely within the output pixel, the value is simply added to the accumulator. Otherwise, if the incoming pixel overlaps two output pixels, the weight for the second pixel is determined. This becomes the new accumulator and the accumulator plus the remainder of the pixel is passed on to the Y mapping. Note that the accumulator has two parts to it: the total accumulated pixel value, and the number of pixels contributed, to enable the value to be normalized.

$$\mathbf{Acc}[Total, Area] = \begin{cases} \mathbf{Acc} + [Pixel, 1] & \text{if } x_u - map_r(x_f) < 1 \\ w_x \times [Pixel, 1] & \text{otherwise} \end{cases} \quad (20)$$

where the weight is

$$w_x = x_u + 1 - map_r(x_f). \quad (21)$$

The completed output is

$$\begin{aligned}\mathbf{Out}_x &= \mathbf{Acc} + (1 - w_x) \times [Pixel, 1] \\ &= \mathbf{Acc} + [Pixel, 1] - w_x \times [Pixel, 1]\end{aligned} \quad (22)$$

As a pixel is output, x_f is also incremented, beginning accumulation for the next pixel. Therefore, the previous value of x_f is used to index the row buffer an image in Eqs. (23) to (26).

A similar process is used for the mapping in the Y direction, with the output from the X mapping added to the accumulator for the column. Since all of the columns are processed in parallel, a row buffer is used to hold the accumulators for each of the output columns until sufficient rows have accumulated to complete an output pixel. The accumulation therefore is

$$\mathbf{RowBuffer}(x_f - 1) = \begin{cases} \mathbf{RowBuffer}(x_f - 1) + \mathbf{Out}_x & y_u - map_r(y_f) < 1 \\ w_y \times \mathbf{Out}_x & \text{otherwise} \end{cases}$$

$$(23)$$

where the weight

$$w_y = y_u + 1 - map_r(y_f) \quad (24)$$

is calculated once at the start of the row. The completed output is

$$\begin{aligned}\mathbf{Out}_y &= \mathrm{RowBuffer}(x_f-1)+(1-w_x)\times\mathbf{Out}_x\\ &= \mathrm{RowBuffer}(x_f-1)+\mathbf{Out}_x-w_x\times\mathbf{Out}_x\end{aligned} \quad (25)$$

Before being saved to the image memory, the output is normalized to give the average pixel value:

$$\mathbf{Image}(y_f,x_f-1)=\frac{\mathbf{Out}_y[Total]}{\mathbf{Out}_y[Area]}. \quad (26)$$

The division required for this may be implemented efficiently (Bailey 2006) using about the same logic as that require for a multiplication.

The most expensive operation is the multiplication for the weighting. For grayscale images, three multiplications are required, one for Eq. (20) and two for Eq. (23); a colour image requires seven multiplications. One (or three for colour) division operations are required for normalization. Note that the lookup table for performing the mapping and the logic for calculating the weight could be multiplexed between the rows and columns because the vertical mapping only needs to be accessed once at the start of each row.

4.4 Radial Mapping Implementation

The radial mappings are more complex than the separable mapping because of the dependency between the X and Y components. For the L_∞ mapping, Eq. (18) implies that the maximum dimension can be determined from the mapping as in Fig. 5. This can be clearly seen in the larger images in Fig. 4. The diagonal partitions of the mapping from the input to the foveal image mean that the reverse mapping used for the separable map provides insufficient information to correctly map the pixels, and the larger forward mapping must be used.

The two-pass mapping also cannot be used directly, because it requires all input pixels that map to a given column to be treated the same. The diagonal partitions imply that some of these map to two separate rows. This becomes worse close to the diagonals, especially when the acuity is low (in the periphery when a large fovea is used) giving the input region for a pixel a large aspect ratio. As a result, several output rows may need to be accumulated simultaneously for a given low resolution column. This case was not considered adequately in the original proposal (Bailey and Bouganis 2008) which only allowed two simultaneous output rows to be accumulated per input column.

A compromise for the L_∞ mapping is to make all of the pixels rectangular rather than trapezoidal, as shown in Fig. 6. As these are in the periphery, where the resolution is lower, the modified geometry is less significant than in the fovea. The pixels along the diagonals in the foveal image, however, can have significantly lower resolution than other pixels.

Such a change would allow a two-pass circuit similar to Fig. 5 to be used. A reverse mapping could still be used, but would also need to include the width of the rectangles as well to enable the output pixel to be determined. Extra logic and an additional buffer would be required to determine the output pixel, and hence the

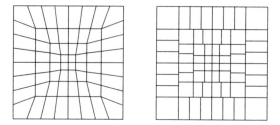

Fig. 6. Rectangularising the partitions for the L_∞ mapping. The original radial mapping is on the left, and the partitions are made rectangular in the mapping on the right.

weights. The buffer would enable incremental calculations to be used to accumulate the rectangle widths to avoid the need for an additional multiplication.

These techniques cannot easily be adapted for the L_2 mapping. Fig. 7 shows a scheme for the direct mapping of each input pixel using a radial mapping. When an input pixel spans multiple output pixels, it is necessary to determine the proportion of the input pixel that maps to each output pixel in order to apportion the pixel value accordingly. The geometric calculations can be quite complex, so a relatively simple approximation is to map the midpoints of the top, bottom, left, and right edges of each pixel. The right edge of one pixel becomes the left edge of the next pixel, and the bottom edge can be cached in a buffer to give the top edge in the next row. Therefore, only two points need to be transformed. A radial mapping requires that the input coordinates be scaled radially, so rather than store the mapping, the forward magnification function of Eq. (3) is used to scale the input coordinates to give the output coordinates. If (x_u, y_u) is the top left corner of an input pixel then the edges are transformed as

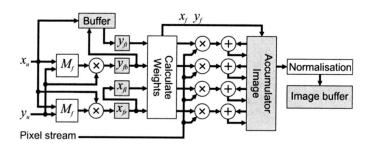

Fig. 7. Structure for radial foveal mappings

$$\begin{aligned} x_{fr} &= (x_u + 1) M_f \left(\sqrt{(x_u+1)^2 + (y_u+\tfrac{1}{2})^2} \right) \\ y_{fb} &= (y_u + 1) M_f \left(\sqrt{(x_u+\tfrac{1}{2})^2 + (y_u+1)^2} \right) \end{aligned} \quad (27)$$

The squaring in Eq. (27) may be performed without multiplication since x_u increments from pixel to pixel and exploiting the fact that

$$(a+1)^2 = a^2 + 2a + 1. \tag{28}$$

Also, the square root is not necessary if the magnification function is stored in a lookup table and is indexed by u^2 rather than u, i.e. $M_f(u^2)$ (Gribbon et al. 2003).

If the integer parts of the pixel edges are the same then the weight is 1, otherwise the weight for apportioning the pixel horizontally is

$$w_x = \frac{\text{fract}(x_{fr})}{x_{fr} - x_{fl}}, \tag{29}$$

and similarly for the vertical weight. The input pixel value is then multiplied by the weights, and both the total and area accumulated for each output pixel. This will require up to four simultaneous accesses to the accumulator image (if the input is spread over four output pixels). This may be achieved by partitioning the accumulator image over four memory blocks, using separate blocks for the odd and even row and column addresses.

After all of the input pixels have been accumulated for an output pixel, it can then be normalized by dividing the total by the area. This gives the output pixel value for the foveal image.

The hardware resource requirements may again be estimated in terms of the number of multiplications and divisions. The magnification lookup tables will need to be interpolated tables, although a multiplication may be avoided with the interpolation by using a bipartite table (Schulte and Stine 1997). If implemented with dual-port memory, only a single magnification table is required. Two multiplications are required to calculate the pixel edges in the foveal image. Two divisions and a multiplication are required to calculate the weights w_x, w_y, and $w_x w_y$. To apportion the input pixel value, only three multiplications are required because the sum of the weights is one. Finally one division is required to normalize the output value. Thus a total of six (twelve for colour) multiplications and three (five for colour) divisions are required by this circuit. The intermediate memory requirements for this mapping are also larger. The buffer for y_{fb} must be the width of the input image, and an accumulator image is required.

5 Discussion and Conclusions

The mapping logic is the smallest for the separable mapping, although only slightly more is needed for a two-pass L_∞ mapping. The L_2 mapping is the most complex. Of the mappings explored, the two-pass L_∞ mapping provides the best compromise in terms of resource requirements and uniformity of pixel size.

With all of the mapping structures presented here, the mapping may be specified in a lookup table. This gives significant flexibility, as changing the mapping requires only changing the contents of the lookup table. Multiple foveal maps may be predefined and selected depending on the application. If the separable mapping is used, the mapping could even be adjusted dynamically between frames, given a suitable mechanism for defining a new mapping. This is more complex for the radial mappings, as their efficient implementation require more details to be precalculated.

The low storage of the warped images and modest resource requirements for implementing the mapping also mean that multiple warp engines could be implemented in parallel if the processing requires different resolutions for different steps.

The design presented in this chapter meets the design requirements for active foveal vision systems identified in the introduction. The position of the fovea may be repositioned dynamically from frame to frame by using a windowed access from a CMOS sensor. An embedded system, consisting of a sensor and FPGA is possible, enabling high speed, active vision to be implemented without a mechanical pan-tilt platform. It is shown that a foveal mapping can reduce the image size significantly, reducing the storage and potential processing time of the output image. In the examples presented here, a 512×512 is reduced to 64×64 – a data reduction factor of 64. The mapping is performed on the data as it is streamed from the sensor, reducing the latency and enabling real-time operation.

Further research is required on the imaging algorithms for processing the foveated images. In particular, saliency detection algorithms that operate on the foveated images require further study. A range of both tracking and recognition algorithms need to be explored to determine the benefits and limitations of the mappings proposed here. Schemes for dynamically adapting and optimizing the mapping function for an application require investigation.

References

Altera, Stratix IV Device Handbook, vol. SIV5V1-2.0. Altera Corporation (2008)

Arribas, P.C., Maciá, F.M.H.: FPGA implementation of a log-polar algorithm for real time applications. In: Conference on Design of Circuits and Integrated Systems, Mallorca, Spain, pp. 63–68 (1999)

Bailey, D.G.: Space efficient division on FPGAs. In: Electronics New Zealand Conference (EnzCon 2006), Christchurch, New Zealand, pp. 206–211 (2006)

Bailey, D.G., Bouganis, C.S.: Reconfigurable foveated active vision system. In: International Conference on Sensing Technology, Tainan, Taiwan, pp. 162–169 (2008)

Bailey, D.G., Bouganis, C.S.: Tracking performance of a foveated vision system. In: International Conference on Autonomous Robots and Agents (ICARA 2009), Wellington, New Zealand, pp. 414–419 (2009)

Bandera, C., Scott, P.D.: Foveal machine vision systems. In: IEEE International Conference on Systems, Man and Cybernetics, Cambridge, Massachusetts, USA, vol. 2, pp. 596–599 (1989)

Bayer, B.E.: Colour filter array. United States of America patent 3971065 (1976)

Camacho, P., Arrebola, F., Sadoval, F.: Shifted fovea multiresolution geometries. In: International Conference on Image Processing, Lausanne, Switzerland, vol. 1, pp. 307–310 (1996)

Camacho, P., Arrebola, F., Sadoval, F.: Multiresolution sensors with adaptive structure. In: 24th Annual Conference of the IEEE Industrial Electronics Society (IECON 1998), Aachen, Germany, vol. 2, pp. 1230–1235 (1998)

Catmull, E., Smith, A.R.: 3-D transformations of images in scaline order. ACM SIGGRAPH Computer Graphics 14(3), 279–285 (1980)

Cui, Y., Samarasekera, S., Huang, Q., Greiffenhagen, M.: Indoor monitoring via the collaboration between a peripheral sensor and a foveal sensor. In: 1998 IEEE Workshop on Visual Surveillance, Bombay, India, pp. 2–9 (1998)

Gribbon, K.T., Johnston, C.T., Bailey, D.G.: A real-time FPGA implementation of a barrel distortion correction algorithm with bilinear interpolation. In: Image and Vision Computing New Zealand 2003, Palmerston North, New Zealand, pp. 408–413 (2003)

Hsia, S.C.: Fast high-quality color-filter-array interpolation method for digital camera systems. Journal of Electronic Imaging 13(1), 244–247 (2004)

Hua, H., Liu, S.: Dual-sensor foveated imaging system. Applied Optics 47(3), 317–327 (2008)

Itti, L., Koch, C., Niebur, E.: A model of saliency-based visual attention for rapid scene analysis. IEEE Transactions on Pattern Analysis and Machine Intelligence 20(11), 1254–1259 (1998)

Kuniyoshi, Y., Kita, N., Sugimoto, K., Nakamura, S., Suehiro, T.: A foveated wide angle lens for active vision. In: IEEE International Conference on Robotics and Automation, Nagoya, Japan, vol. 3, pp. 2982–2988 (1995)

Liu, Y., Bouganis, C.S., Cheung, P.Y.K.: A spatio-temporal saliency framework. In: IEEE International Conference on Image Processing, Atlanta, Georgia, USA, pp. 437–440 (2006)

Martinez, J., Altamirano, L.: FPGA-based pipeline architecture to transform cartesian images into foveal images by using a new foveation approach. In: IEEE International Conference on Reconfigurable Computing and FPGA's, San Luis Potosi, Mexico, pp. 1–10 (2006)

Ovod, V.I., Baxter, C.R., Massie, M.A., McCarley, P.L.: Advanced image processing package for FPGA-based re-programmable miniature electronics. In: Infrared Technology and Applications XXXI, Orlando, Florida, USA. SPIE, vol. 5783, pp. 304–315 (2005)

Schulte, M.J., Stine, J.E.: Symmetric bipartite tables for accurate function approximation. In: 13th IEEE Symposium on Computer Arithmetic, Asilomar, California, USA, pp. 175–183 (1997)

Traver, V.J., Pla, F.: Designing the lattice for log-polar images. In: Nyström, I., Sanniti di Baja, G., Svensson, S. (eds.) DGCI 2003. LNCS, vol. 2886, pp. 164–173. Springer, Heidelberg (2003)

Vogelstein, R.J., Mallik, U., Culurciello, E., Etienne-Cummings, R., Cauwenberghs, G.: Spatial acuity modulation of an address-event imager. In: 11th IEEE International Conference on Electronics, Circuits and Systems (ICECS 2004), Tel-Aviv, Israel, pp. 207–210 (2004)

Wang, Z., Bovik, A.C.: Embedded foveation image coding. IEEE Transactions on Image Processing 10(10), 1397–1410 (2001)

Wilson, J.C., Hodgson, R.M.: A pattern recognition system based on models of aspects of the human visual system. In: International Conference on Image Processing and its Applications, Maastricht, Netherlands, pp. 258–261 (1992a)

Wilson, J.C., Hodgson, R.M.: Log-polar mapping applied to pattern representation and recognition. In: Computer Vision and Image Processing, pp. 245–277. Academic Press, London (1992b)

Wolberg, G.: Digital image warping. IEEE Computer Society Press, Los Alamitos (1990)

Xilinx, Virtex-5 FPGA User Guide, vol. UG190 (v4.4). Xilinx Inc. (2008)

Xue, Y., Morrell, D.: Adaptive foveal sensor for target tracking. In: Thirty-Sixth Asilomar Conference on Signals, Systems and Computers, Pacific Grove, California, USA, vol. 1, pp. 848–852 (2002)

Development of a Real-Time Full-Field Range Imaging System

A.P.P. Jongenelen[1], A.D. Payne[2], D.A. Carnegie[1], A.A. Dorrington[2], and M.J. Cree[2]

[1] School of Engineering and Computer Science,
 Victoria University of Wellington, Wellington, New Zealand
 `adrian.jongenelen@vuw.ac.nz`
[2] Department of Engineering,
 University of Waikato, Hamilton, New Zealand

Abstract. This article describes the development of a full-field range imaging system employing a high frequency amplitude modulated light source and image sensor. Depth images are produced at video frame rates in which each pixel in the image represents distance from the sensor to objects in the scene.

The various hardware subsystems are described as are the details about the firmware and software implementation for processing the images in real-time. The system is flexible in that precision can be traded off for decreased acquisition time. Results are reported to illustrate this versatility for both high-speed (reduced precision) and high-precision operating modes.

1 Introduction

Range Imaging is a field of electronics where a digital image of a scene is acquired, and every pixel in that digital representation of the scene records both the intensity and distance from the imaging system to the object. The technique described here achieves this by actively illuminating the scene and measuring the time-of-flight (TOF) for light to reflect back to an image sensor [1-5]. Alternative techniques for forming range or depth data include stereo vision, active triangulation, patterned light, point scanners and pulsed TOF [6-8].

Applications of range imaging can be found in mobile robotics, machine vision, 3D modelling, facial recognition and security systems.

This article describes the development of a full-field range imaging system assembled from off-the-shelf components. The system employs a modulated light source and shutter set up in a heterodyning configuration [9] as described in Section 2. Sections 3 and 4 introduce the various hardware components of the system and the software and firmware used to process the image sequence in real-time and provide control by the user [10]. A summary of the system is presented in Section 5, illustrating experimental results of captures taken both at high-speed (reduced precision) for a dynamic scene and also with high-precision in a static scene.

2 Time-of-Flight Range Imaging Theory

The full-field range imaging technique described here utilises an Amplitude Modulated Continuous Wave (AMCW) light source and image sensor to indirectly measure the time for light to travel to and from the target scene. In this section we describe two techniques, homodyne modulation and heterodyne modulation, that we employ for indirect acquisition of time-of-flight. In both cases the target scene is illuminated by an AMCW light source with frequency, f_M, typically between 10 and 100 MHz.

2.1 Homodyne Modulation

When illuminated by the AMCW light source, objects in the scene reflect the light back to an image sensor that is amplitude modulated at the same frequency, f_M as the light source. Due to the time taken for the light to travel to and from the scene a phase delay, θ, is evident in the modulation envelope of the received signal.

The modulated sensor effectively multiplies (or mixes) the returned light waveform with the original modulation signal. This is then integrated within the pixel to give an averaged intensity value relative to the phase offset. Figure 1 shows example waveforms with two objects at different distances, giving phase shifts θ_1 and θ_2 for the returned light which produce two different intensities I_1 and I_2.

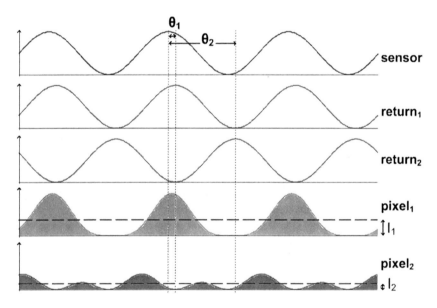

Fig. 1. Demonstration of pixel intensity as a function of phase delay for light returned from two objects at different distances. The reflected signals are mixed with a copy of the original modulation waveform producing pixel$_1$ and pixel$_2$. The total shaded area is proportional to the returned intensity of the pixels, I_1 and I_2.

From a single frame it is impossible to know if reduced pixel intensity is a result of the phase offset of a distant object or due to other parameters such as object reflectance or simply the returned light intensity dropping with distance by the relationship $1/d^2$. To resolve this ambiguity, N frames are taken with an artificial phase step introduced in the modulated sensor signal relative to the light signal incrementing by $2\pi/N$ radians between each frame. The phase delay can now be calculated using a single bin Discrete Fourier Transform as [11]

$$\theta = \arctan\left(\frac{\sum I_i \cos\left(\frac{2\pi i}{N}\right)}{\sum I_i \sin\left(\frac{2\pi i}{N}\right)}\right) \quad (2.2)$$

where I_i is the pixel intensity for the ith sample. Typically $N = 4$ as this simplifies the calculation to [12]

$$\theta = \arctan\left(\frac{I_0 - I_2}{I_1 - I_3}\right) \quad (2.3)$$

removing the need for storing or calculating the sine and cosine values. It is also simpler for the electronics to introduce the artificial phase shifts $\pi/2$, π and $3\pi/2$ radians by deriving the signal from a master clock operating at twice the modulation frequency.

From the phase shift, θ, the distance can then be calculated as

$$d = \frac{c\theta}{4\pi f_M} \quad (2.4.)$$

where c is the speed of light and f_M is the modulation frequency.

2.2 Heterodyne Modulation

Heterodyne modulation differs from homodyne modulation in that the illumination and image sensor are modulated at slightly different frequencies. The illumination is modulated at the base frequency, f_M, and the sensor is modulated at $f_M + f_B$ where the beat frequency, f_B is lower than the sampling rate of the system. The sampling rate, f_S, is ideally an integer multiple of f_B as calculated by Equation 2.5.

$$f_B = \frac{f_S}{N} \quad (2.5.)$$

The effective result is a continuous phase shift rather than the discrete phase steps between samples of the homodyne method. The phase shift during the sensor integration period reduces the signal amplitude, in particular attenuating higher frequency harmonic components which can contaminate the phase measurement, producing nonlinear range measurements. Through careful selection of N and the length of the integration time, the heterodyne method minimizes the effects of harmonics better than is normally possible with a homodyne system. This enhances measurement accuracy [13].

Another advantage is the ability of the system to easily alter the value of N from three to several hundred simply by changing the beat frequency. This allows the user to select between high-speed, reduced-precision measurements with few frames (N small) or high-precision measurements with many frames (N large).

3 Hardware Components

A working prototype of a heterodyne ranging system has been constructed at the University of Waikato (Hamilton, New Zealand) from off-the-shelf components. This consists of an illumination subsystem, industrial digital video camera, image intensifier unit to provide the high-speed sensor modulation, high-frequency signal generation and support electronics for real-time range processing and control. Fig. 2 shows an overview of the hardware configuration.

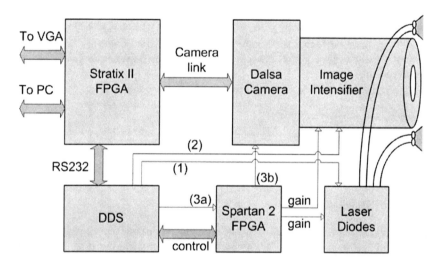

Fig. 2. Heterodyne Ranger system hardware configuration with three modulation outputs identified as (1) f_M, (2) $f_M + f_B$, (3a) mf_S and (3b) f_S

3.1 Camera

A Dalsa Pantera TF 1M60 [14] digital camera is used to record the image sequence. It has a 12-bit ADC and a resolution of up to 1024 × 1024 pixels at a frame rate of up to 60 Hz. For this application the camera is set to operate in 8 × 8 binning mode improving the per pixel signal to noise ratio and increasing the maximum frame rate to 220 Hz. The reduced 128 × 128 resolution also eases the real-time processing memory requirements. The video data stream is provided from the camera via the industry standard CameraLink interface [15].

3.2 Image Intensifier

A Photek 25 mm single microchannel plate (MCP) image intensifier is used to provide high-speed gain modulation of the received light. Gated image intensifier applications typically pulse the photocathode voltage with 250 V_{p-p} at low repetition rates to provide good contrast between on and off shutter states, but this is not feasible for continuous modulation at frequencies up to 100 MHz. Instead the photocathode is modulated with a 50 V_{p-p} amplitude signal that can be switched at the desired high frequencies, but at the expense of reduced gain and reduced image resolution [16]. In this particular application, the reduction in imaging resolution does not impede system performance because the camera is also operated in a reduced resolution mode (8 × 8 binning mode).

3.3 Laser Illumination

Illumination is provided by a bank of four Mitsubishi ML120G21 658 nm wavelength laser diodes. These diodes are fibre-optically coupled to an illumination mounting ring surrounding the lens of the image intensifier. This enables the light source to be treated as having originated co-axially from the same axis as the camera lens and reduces the effect of shadowing. The fibre-optics also act as a mode scrambler providing a circular illumination pattern, with divergence controlled by small lenses mounted at the end of each fibre. Maximum optical output power per diode is 80 mW which is dispersed to illuminate the whole scene.

3.4 Modulation Signal Generation

The heterodyne ranging technique requires two high frequency signals with a very stable small difference in frequency. In order to calculate absolute range values, an additional third signal is required to synchronise the camera with the beat signal so that samples are taken at known phase offsets.

A circuit board containing three Analog Devices AD9952 Direct Digital Synthesizer (DDS) chips is used to meet these requirements [17]. The board uses a 20 MHz temperature-controlled oscillator which each DDS IC multiplies internally to generate a 400 MHz system clock. The DDS ICs also have the ability to synchronise their 400 MHz system clocks to remove any phase offset between the outputs.

The output signals are sinusoidal with each frequency programmed by a 32-bit Frequency Tuning Word, *FTW*, via the SPI interface of an Atmel 89LS8252 microcontroller. The output frequency, f, is calculated as

$$f = \frac{f_{SYS} FTW}{2^{32}} \qquad (3.1.)$$

where f_{SYS} is the 400 MHz DDS System clock frequency [18]. This implies a frequency resolution of 0.093 Hz corresponding to the minimal increment of one in the *FTW*.

The three signals generated by the DDS board are used for:

1) the laser diode illumination, f_M,
2) the modulation of the image intensifier, $f_M + f_B$ and
3) the camera frame trigger multiplied by a constant, mf_S.

The DDS outputs derived from the same master clock ensures appropriate frequency stability between the beat signal, f_B, and the camera trigger mf_S.

A Xilinx Spartan 2 Field Programmable Gate Array (FPGA) receives the sinusoidal mf_S signal, and using a digital counter divides the signal down to produce a CMOS signal of f_S used to trigger the camera. This produces less jitter than that produced by the alternative of passing the less than 200 Hz sinusoidal signal directly to a comparator to effect a sinusoidal to digital conversion.

3.5 Control Hardware

The Spartan 2 board controls the analogue gain of the image intensifier and the laser diodes using a pair of Digital to Analogue Converter ICs. The gain control of the laser diodes is used to ensure that they are slowly ramped on over a period of several minutes to protect them from over-power damage before they reach the stable operating temperature (operating in constant current mode). Controlling the gain of the image intensifier ensures the full dynamic range of the camera sensor is utilised while preventing pixel saturation.

Top level control of the system is provided through an Altera Stratix II FPGA resident on an Altera Nios II Development Kit connected via a JTAG connection to a PC. Control instructions are processed by a Nios II soft processor core and written to either special function registers within the Stratix II FPGA or passed on via RS232 to the microcontroller of the DDS board. The Spartan 2 board is controlled through an 8-bit port of the DDS board microcontroller.

3.6 Processing Hardware

In addition to providing top level control of the various hardware components of the system, the Stratix II FPGA is used to process the image sequence and calculate the range data using Equation 2.2.

Frames are directly transferred from the Dalsa camera over the CameraLink interface to the FPGA. A daughter board utilising a National Semiconductor DS90CR286 ChannelLink Receiver IC is used to convert from the serial LVDS CameraLink format to a parallel CMOS format more acceptable for the general purpose I/O pins of the FPGA.

The camera has two data taps, each streaming pixel information at a rate of 40 MHz. The images from the camera are processed internally within the FPGA and temporarily stored in on-chip block RAM ready to be presented to the user.

3.7 User Interfaces

The system is controlled through the Altera Nios II terminal program, which communicates with the Nios II processor through a JTAG interface. This is used to set system parameters including frame rate, modulation frequency, beat frequency, number of frames per beat, image intensifier gain, laser diode gain and various other special function registers.

For real-time range data display, a daughter board with a Texas Instruments THS8134 Triple 8-bit DAC IC is used to drive a standard VGA monitor. For longer term storage and higher level processing the range data are transferred to a PC using a SMSC LAN91C11 Ethernet MAC/PHY chip resident on the Nios II Development Kit. The embedded Nios II software handles the fetching of the temporarily stored range data from the block RAM and sends it out through a TCP connection to a host PC.

4 Software and Firmware Components

There are three distinct sets of software involved in operating the system, namely:

1) Firmware for the real-time calculation of range data on the Stratix II FPGA,
2) Software on the Nios II processor and
3) Host software on the PC.

4.1 Range Processing Firmware

The data stream coming from the camera consists of two pixel channels at 40 MHz each. It is very difficult to calculate the phase shift of each pixel using Equation 2.2 in real-time with current sequential processors, whereas the configurable hardware of the FPGA is ideally suited to this task as it can perform a large number of basic operations in parallel at the same clock rate as the incoming pixels.

Figure 3 shows an overview of the FPGA firmware. Control signals originating from the user-operated PC are shown in black and the image data path originating from the camera is shown in white.

The range processing firmware can be divided into four sections: The CameraLink interface to the Dalsa Camera, the multiply and accumulate calculation, the arc tangent calculation, and the interface to the Nios II processor. All of the range calculation firmware is instantiated twice to independently process each of the two output taps of the camera. The process is fully pipelined with a latency of 24 clock cycles, and a theoretical maximum frequency of 144 MHz. A 125 MHz system clock is used for all elements internal to the FPGA.

The firmware for this project is designed in VHDL and simulated using Aldec's Riviera Pro software. Altera's Quartus II software is used to synthesise the design and download it to the board.

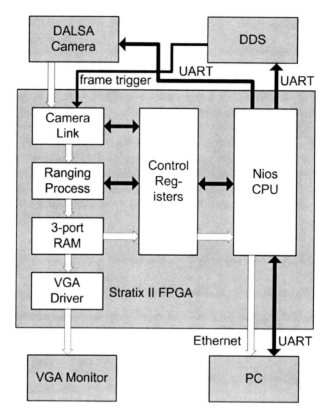

Fig. 3. Top level overview of FPGA configuration. The data path of captured and processed image frames are shown as white arrows. Control signals are shown as black arrows.

The CameraLink Interface

The interface to the Dalsa camera consists of 27 parallel inputs with a 40 MHz clock input for providing the pixel data, one output for the frame trigger of the camera, and one line in each direction for UART serial communication [15].

The frame trigger is provided directly from the Spartan 2 board and simply passes through the interface to the camera. The serial communication lines are handled by the Nios II processor.

All of the 40 MHz data lines are sampled at the 125 MHz FPGA system clock. The *lval* and *fval* control inputs are used to specify when a valid line and frame of data are being received respectively. These are monitored by a state machine as shown in Figure 4 to give feedback to the system when a new frame is complete.

The state machine first ensures that it is aware of which part of the transfer sequence is currently underway, either during integration while the frame trigger is high or during readout while *fval* is high. While the frame and line valid signals are both high, valid pixel data are sampled on the falling edge of the 40 MHz clock. Once the frame and line valid signals are both low the readout is complete

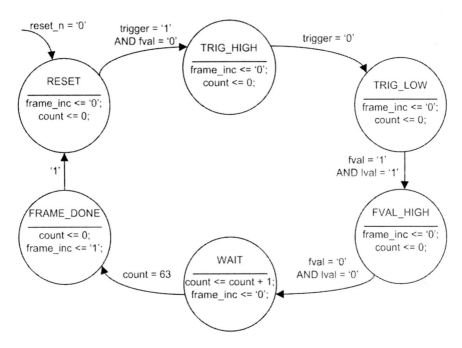

Fig. 4. CameraLink finite state machine for tracking incoming frames. Data readout is initiated by a falling edge of *trigger* and is read from the camera during the *FVAL_HIGH* state. A high pulse of *frame_inc* is generated a short time after readout is complete to prepare the system for the next frame.

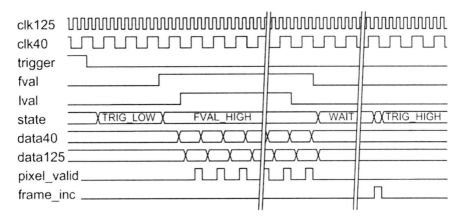

Fig. 5. Timing of the signals internal to the CameraLink interface. Readout from the camera is initiated by a falling edge of the trigger signal. Data is transferred from the camera synchronous to a 40 MHz clock, and is sampled within the FPGA using a 125 MHz clock. When *fval* and *lval* are high, the falling edge of the 40 MHz clock generates a high pulse of *pixel_valid*.

and a counter is started to introduce a delay before the frame increment signal is pulsed high. This allows the processing electronics time to complete before various registers and look-up-tables are updated for the next frame. Figure 5 shows the timing of the signals within the CameraLink interface.

Multiply and Accumulate Firmware

This section details the calculation of summations in Equation 2.2. The equation is first broken down into a series of basic operations to calculate the numerator and denominator (termed as the real and imaginary parts respectively) using intermediate values as labelled in Equation 4.1. The frame index, i, refers to the current frame number between 0 and N-1 which is incremented with each iteration through the state machine.

$$\begin{aligned}
\text{real\_new} &= \text{raw\_pixel} \cdot \text{cos\_lut}(i) \\
\text{imag\_new} &= \text{raw\_pixel} \cdot \text{sin\_lut}(i) \\
\text{real\_acc} &= \text{real\_new} + \text{real\_old} \\
\text{imag\_acc} &= \text{imag\_new} + \text{imag\_old}
\end{aligned} \quad (4.1.)$$

The sine and cosine operations are pre-calculated based on the number of frames per beat, N, and stored in the look-up-tables (LUTs) *sin_lut* and *cos_lut* respectively. These values can be changed by the user through control registers when the desired value of N is changed. The values for *real_old* and *imag_old* are the values of *real_acc* and *imag_acc* from the previous frame which have been stored in block RAM.

Each operation is implemented as a dedicated block of hardware as shown by Figure 6. Operations are pipelined such that the calculation for any newly received pixel can begin before the calculation of the previous pixel is fully completed.

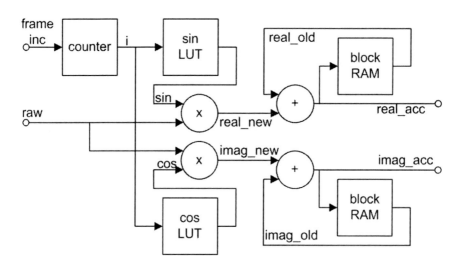

Fig. 6. Firmware block diagram implementing equation 4.1. Each operation has a dedicated block of hardware and the whole calculation is fully pipelined.

Not shown is the circuitry to reset the value of i to 0 after N frames have been processed. At the start of each beat sequence ($i = 0$) the values of *real_old* and *imag_old* are taken as 0 to clear the value of the accumulator. The block RAM is dual ported, which allows values to be written and read simultaneously.

For real-time processing, the amount of on-chip block RAM is the greatest limiting factor in the resolution of the range images, because every pixel of the image requires enough storage for both the real and imaginary accumulated values. In the current implementation using an Altera Stratix II EP2S60 FPGA this restricts the resolution to 128×128 pixels.

Arc Tangent Calculation

The arc tangent is approximated using a look-up-table (LUT). Figure 7 shows how the symmetric nature of the arc tangent function is utilised to reduce the RAM requirements of the LUT to one eighth of the full cycle. The sector, out of the eight possible sectors, of the final result is determined by the sign and magnitude of the

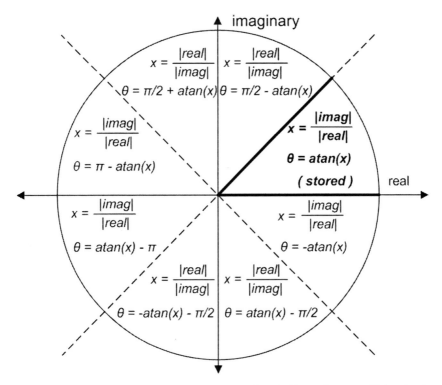

Fig. 7. Calculation of arc tangent using one eighth of the LUT requirements. Using the sign and relative magnitude of the real and imaginary inputs determines which region the result will be in. The index into the LUT is the scaled result of the smallest of the two divided by the largest. This is added to or subtracted from a constant to find the final arc tangent approximation.

real and imaginary components. The division is calculated and used to index the LUT, and the intermediate result is added to or subtracted from a constant to give the final result.

The advantage of calculating the arc tangent in this way is that the result of the division is always a positive number between 0 and 1. This scales easily as an index into the look-up-table. With a 16-bit by 1024 point LUT, the arc tangent can be approximated to better than 0.8 milliradians.

Interface to Nios II processor

The output frame from the arc tangent function is stored in a portion of triple port block RAM. The three ports are designated as one write-only port for the result of the arc tangent calculator and two read-only ports: one for the VGA display driver and the other for the Nios II processor.

The Nios II processor addresses 32k memory locations in the hardware. The lower 16k reference the output RAM, and the upper 16k reference special function control registers. These registers control everything within the Stratix II FPGA such as sine and cosine LUT values, frames per beat and a synchronous reset. Two control registers are set up in relation to the transfer of data to the processor:

- OR_RDY: a 1-bit read-only register which is set high by hardware when a processed frame is available in the output RAM.
- OR_ACK: a 1-bit write-only register which is used to signal to the hardware that reading of the output RAM is complete.

While the OR_RDY bit is high, writing into the output RAM is disabled. This ensures that a frame being transferred out is not inadvertently overwritten.

4.2 Nios II Software

The Nios II processor is a reconfigurable microprocessor core provided by Altera. The features of the processor can be selected using the SOPC builder available with the Quartus II software.

For this application, a processor has been designed with the following features:

- Nios II Processor with 4 kB instruction cache
- 32 MB DDR SDRAM
- Three serial UART interfaces for
    - DDS control
    - Dalsa camera control
    - PC user interface
- LAN91C111 ethernet MAC/PHY
- 16 × 2 character LCD display
- 32-bit parallel interface to on-chip ranger hardware

The Nios II processor is programmed in C using the Nios II IDE software. The main tasks performed by the processor are to:

1) Parse text commands from the user and either a) read or write on-chip control registers b) forward the message on to the DDS board microcontroller c) forward the message on to the Dalsa camera
2) Respond to the OR_RDY signal generated by the processing hardware, and
3) Handle the TCP connection for transferring frames to the PC.

A multi-threaded program is used for these tasks with the TCP packet handler having top priority. While the multi-threaded environment does add extra complexity and overhead to the code, an Altera example design with ethernet and TCP/IP configuration functionality was simply modified to include the additional required tasks.

4.3 PC Software

Software on the PC is programmed in C# using Microsoft Visual Studio. A basic terminal program establishes a TCP connection with the Nios II processor of the Stratix II FPGA, listening for incoming packets and writing them to a binary file.

The Nios II IDE software also includes a terminal for communicating with the Nios II processor through JTAG. This interface is used for sending text commands to the Nios II processor for setting various control parameters.

5 Results and Summary

The following results give an indication of ranging performance for two operating modes of the system: a capture of a dynamic scene for high-speed range images at the expense of depth precision, and a capture of a static scene for high precision measurements.

5.1 Test Capture of a Dynamic Scene

To demonstrate the operation of the range imaging system with video-rate output a scene has been set up with a number of moving objects. These are a pendulum swinging in a circular motion, a bear figurine rotating on a turntable and a roll of paper towels rolling down a ramp towards the camera. System parameters for this capture are $f_M = 40$ MHz, $f_S = 125$ Hz and $N = 5$, resulting in 25 range images per second.

Figure 8 shows the first frame of the sequence with a group of labelled test pixels. These are locations in the capture showing 1) the area above the bear imaging a cardboard background, 2) a region of the far wall of the room and 3) to 5) the path of the rolling paper towels. The rotating bear and the pendulum, which is the dark spot to the right of the bear are not analysed in this capture.

Figure 9 shows a plot of the test pixels as their measured range changes throughout the 40 range image capture. As expected for stationary objects, the range at pixels 1 and 2 does not change throughout the capture and give an

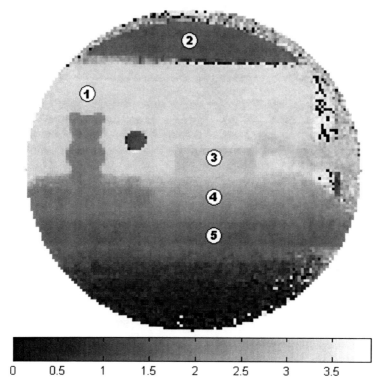

Fig. 8. First range image frame of dynamic test capture with identified test pixels: 1) cardboard background, 2) far wall, 3) to 5) pixels tracking the rolling paper towels. Grey scale pixel intensity represents range from 0 to 3.75 m.

indication of the precision of the system in this operating mode. Test pixel 1 has a mean of 3.11 m and standard deviation of 4.0 cm and test pixel 2 has a mean of 1.27 m and standard deviation of 3.6 cm.

Although test pixel 2 represents the far wall, it is measured to be a distance similar to pixel 5. This is a consequence of phase wrapping where objects beyond the unambiguous range of $c/2f_M = 3.75$ m are measured incorrectly. It is possible to correct for this ambiguity [9] but this functionality is not currently incorporated into this system.

The plot also shows an interconnecting slope between test pixels 3 to 5 where the paper towels roll towards the camera through these pixels. For example, test pixel 4 is initially returning the range of a location on the ramp at approximately 2.4 m. Between frames 17 and 24 the paper towels roll towards the camera through the location imaged by this pixel evident as a decreasing range measurement. After the paper towels have gone past the location of pixel 4 the value returns to the initial range of 2.4 m.

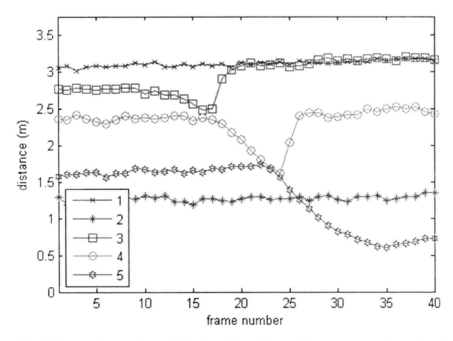

Fig. 9. Measured range of test pixels throughout 40 range image sequence: 1) cardboard background, 2) far wall, 3) to 5) points on the rolling paper towels

5.2 Test Capture of a Static Scene

Figure 10 shows a three-dimensional mesh plot reconstructed from range image data. This capture was taken over 10 seconds with $f_M = 80$ MHz, $f_S = 29$ and $N = 29$ Hz. The combination of large integration time, the high number of frames per beat and averaging over 10 beat cycles results in a range image with a standard deviation of less than 1 mm.

5.3 Summary

This article has described the construction of a heterodyne range imaging system with video-rate depth image output. The system has the following parameters:

- x-y resolution of 128 × 128 pixels
- variable output frame rate with real-time display up to 60 Hz
- variable frames per beat, N
- variable modulation frequency, f_M, up to 80 MHz
- data capture using a common ethernet interface
- system parameters adjustable through a command line interface

With these easily adjustable system parameters it is possible to configure the system to meet the needs of a wide variety of applications from 3D modeling to mobile robotics.

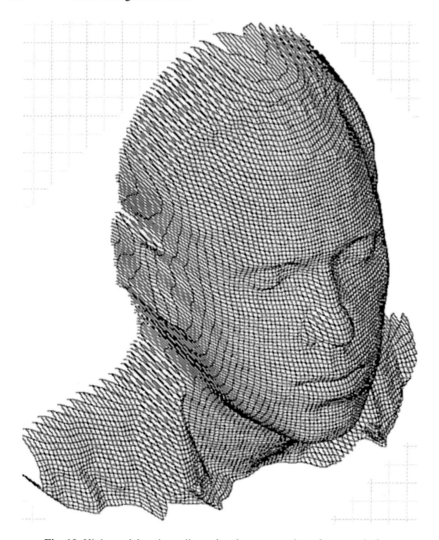

Fig. 10. High precision three-dimensional reconstruction of a person's face

References

[1] Christie, S., et al.: Design and development of a multi-detecting two-dimensional ranging sensor. Measurement Science and Technology 6, 1301–1308 (1995)
[2] Kawahito, S., et al.: A CMOS Time-of-Flight range image sensor with Gates-on-Field-Oxide structure. IEEE Sensors 7, 1578–1586 (2007)
[3] Lange, R., Seitz, P.: Seeing distances – a fast time-of-flight 3D camera. Sensor Review 20, 212–217 (2000)
[4] Gulden, P., et al.: Novel opportunities for optical level gauging and 3-D imaging with the Photoelectronic Mixing Device. IEEE Transactions on Instrumentation and Measurement 51, 679–684 (2002)

[5] Buttgen, B., et al.: High-speed and high-sensitive demodulation pixel for 3D-imaging. In: Proc. SPIE – Three-dimensional image capture and applications, vol. 7 (2006)
[6] Blais, F.: Review of 20 years of range sensor development. Journal of Electronic Imaging 13, 231–243 (2004)
[7] Besl, P.J.: Active, Optical range imaging sensors. Machine Vision and Applications 1, 127–152 (1988)
[8] Sato, K.: Range imaging based on moving pattern light and spatio-temporal matched filter. In: International Conference on Image Processing, vol. 1, pp. 33–36 (1996)
[9] Dorrington, A.A., et al.: Achieving sub-millimetre precision with a solid-state full-field heterodyning range imaging camera. Measurement Science and Technology 18, 2809–2816 (2007)
[10] Jongenelen, A.P.P., et al.: Heterodyne range imaging in real-time. In: Proc. International Conference on Sensing Technology, Tainan, vol. 3, pp. 57–62 (2008)
[11] O'Shea, P.: Phase Measurement. In: Webster, J.G. (ed.) Electrical Measurement, Signal Processing and Displays, pp. 28–41. CRC Press, Boca Raton (2003)
[12] Lange, R., Seitz, P.: Solid-state time-of-flight range camera. IEEE Journal of Quantum Electronics 37, 390–397 (2001)
[13] Dorrington, A.A., et al.: Video-rate or high-precision: a flexible range imaging camera. In: Proc. SPIE Image Processing: Machine Vision Applications, vol. 6813 (2008)
[14] DALSA, Pantera TF 1M60 and 1M30 User's Manual and Reference (2004), http://www.dalsa.com
[15] PULNiX America Inc., Specifications of the Camera Link Interface Standard for Digital Cameras and Frame Grabbers (2000), http://www.imagelabs.com
[16] Payne, A.D., et al.: Image intensifier characterisation. In: Proc. Image and Vision Computing, New Zealand, pp. 487–492 (2006)
[17] Payne, A.D., et al.: A synchronized Direct Digital Synthesiser. In: International Conference on Sensing Technology, Palmerston North, pp. 174–179 (2005)
[18] Analog Devices, 400 MSPS 14-Bit, 1.8 V CMOS Direct Digital Synthesizer AD9952 Datasheet (2009), http://www.analog.com
[19] Altera Corporation, Stratix II Device Handbook, vol. 1 (2007), http://www.altera.com

Development of a Stereo Image Distribution System

Junichi Takeno, Toshihiro Enaka, and Hirofuji Sato
Meiji University Graduate School,
Kawasaki-shi, Japan

Abstract. The authors have developed a remote stereo vision system (RSVS) in which the right and left images captured by stereo cameras are synthesized into single image information, compressed as MPEG-2 data, and transmitted to remote observers over the Internet. This paper presents a stereo image distribution system featuring DVD-class image quality and a new image capture method that relatively reduces the visual strain on the observer inherent in conventional stereo camera capture methods. The image quality of the new method is approximately 13 Mbps. With conventional image capture methods using two stereo cameras arranged in parallel, objects located near to the cameras have a larger disparity, causing eye strain for the observer. The typical method for countering this problem is to have the optical axes of the right and left cameras intersect each other and adjust the position of the intersection when capturing the images. This technique, generally called the intersection method, is effective for reducing the visual strain on the observer. However, the drawback of this method is that the sense of depth is distorted geometrically because the locus of the same disparity, called the horopter, forms a circle. The authors have developed a new stereo camera image capture method to correct this geometrical distortion while reducing visual strain.

1 Introduction

The power and presence of stereo vision is readily experienced in our daily lives. The advantage of stereo images is that relatively correct distance information, which is not obtainable with monocular images, is obtained directly and visually. This feature has enabled remote surgical operations and other novel techniques to be developed and commercialized in recent years. Thanks to the rapid development of LCDs and other display technologies, shrinking sizes of ICs and other devices, and progress in software technologies, consumers are now able to enjoy video games on general-purpose PCs, and stereo vision is featured on mobile terminals. Household TV sets may also be used to view 3D broadcasts albeit in a simple form. While considering the entertainment value and practicality of stereo images, we have developed a system to distribute stereo images remotely. We have also devised and incorporated into the system a scheme to relatively reduce the visual strain on the observer which is peculiar to stereo vision applications.

2 Developed Sterervision System

The authors have developed three systems as described below. Remote Stereo Vision System: This system makes it possible to distribute stereo images remotely in real time. Remote Control System: This system enables the observer to control the stereo cameras remotely, including aiming and zooming the cameras. Record and Edit System: This system is designed to record stereo images and capture images using the parallel translation method.

Remote Stereo Vision System (RSVS)

The Remote Stereo Vision System is used to display remote stereovision with DVD image quality [1]. The right and left images are encoded as MPEG-2 data and transmitted remotely over the Internet. On the receiver side, the MPEG data are decoded into NTSC signals, and the right and left images are perfectly synchronized using the genlock technique. The right and left images are converted into single double-speed images (120 Hz) on an upconverter for alternate display. The resolution on the alternate display is very sharp at 720 x 480. There is, however, a time delay of one second in the image distribution. The observer views the stereo images using the time-sharing shutter system.

Table 1. System Performance

MPEG compression	MPEG-2
Video pixel count	720x480(Single image equivalent)
Video bit rate	13Mbps
Delay time	Approx. 1 sec

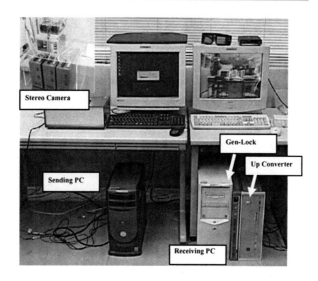

Fig. 1. RSVS SYSTEM

Remote Control System

Remote Control System is designed to control the RSVS cameras remotely. In addition to zoom, focus and white balance setting, the operator can freely adjust the optical axes of the cameras by controlling the servo motors mounted below the stereo cameras. This interactive system enables the operator to easily capture the desired images.

Fig. 2. Front view(left) and Side view(right)

While viewing the stereo vision images distributed by RSVS, the operator located on the image receiving side controls the client software program of the remote controller to adjust the various operating parameters of the stereo cameras such as the convergence angle. This interactive system makes it possible to capture the images that the operator wishes to see.

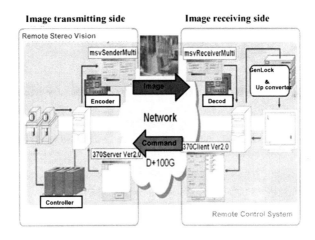

Fig. 3. Correlation between RSVS and Remote Control System

Image Record and Edit System
The authors' present study focuses on the development of a new image capture method to alleviate the visual strain on the observers of stereo images. It would possible to compare multiple image capture techniques effectively if record and reproduce functions were available. With this in mind, we developed the Record and Edit System to determine the effective image capture method. The Record and Edit system makes it possible to record and reproduce stereo images, and capture images of objects using the parallel translation method.

The camera images are retrieved from the encoder board. When recording images with the parallel translation method, the right and left still images are captured from the encoder board, and the amount of movement is calculated by the matching program. The right and left images are recorded simultaneously, and stored on a hard disk. When viewing the stereo images, the stored images are simultaneously reproduced using a decoder software program, and are output to the existing stereovision display interface of the RSVS. By setting the amount of the image movement when encoding, the observer can view the images processed by the parallel translation method.

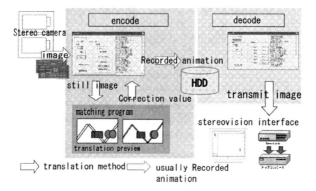

Fig. 4. Structure of Record and Edit System

3 Visual Burden and the New Image Capture Method

The authors' research group has established a high-quality stereo image distribution system using the Internet to address the following problems typically experienced with conventional stereo vision systems.

- Stereoscopic effects are lost depending on the image capture environments and objects.
- Stereoscopic effects are lost depending on the observer even under identical image capture conditions.
- Stereoscopic effects can be lost when the cameras are zoomed.
- Some viewers are affected by the visual burden (eye strain) after viewing stereo vision images for some time.

The present study intends to lessen these stereo vision anomalies and the visual burden as much as possible.

Parallel Method
The parallel method is a popular stereo vision image capture method. Right and left cameras are arranged so that their optical axes are parallel to each other with a certain fixed distance. Parallax decreases as the imaged object moves further in

front of the cameras and eventually disappears at infinity. It is possible to provide images free from geometrical distortion because the locus of the same parallax follows a straight line that is perpendicular to the optical axes of the cameras. It is generally believed that the disadvantage is that the viewing environment is unnatural and visual fatigue sets in for the observer because the focal distance at which the observer's eyes can form sharp images on their retinas is different from the position where the 3D images are produced (convergence distance)[2]. If this is true, the visual burden may be alleviated by adjusting the parallax of the object's images so that the convergence distance and the focal distance are identical.

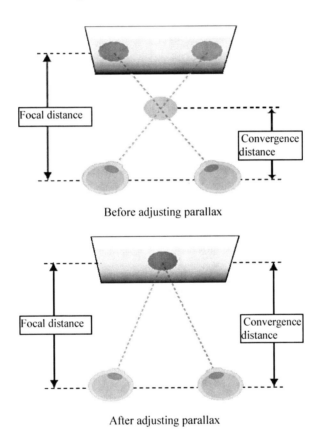

Fig. 5. Alleviation of Visual Burden by Adjusting Parallax

Cross Method
The optical axes of the right and left cameras are crossed on the imaged object. The cross point of the optical axes of the cameras is positioned on the stereo image display plane. It is therefore possible to adjust the parallax by varying the position of the cross point before capturing the image. The disadvantage is that geometrical distortion in the depth is unavoidable because the locus of the same parallax is a circular locus known as the horopter[2][3][4].

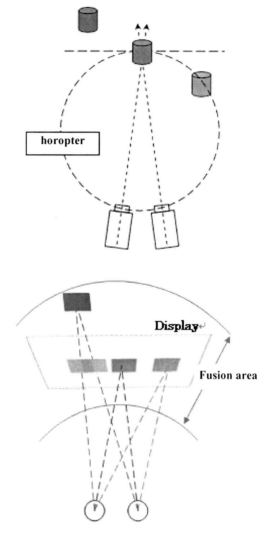

Fig. 6. Camera Layout (above) and Image Forming Position (below) of Cross Method

Translation Method
To realize stereo vision, a correct sense of distance through depth recognition is important, in addition to an enhanced sense of presence.

As such, the fact that the images are geometrically distorted is a critical disadvantage of the cross method. To capture more comfortable-appearing stereo images, we must devise a new image capture technique that features the merits of both the parallel method and the cross method. The authors propose a new image capture method called the translation method.

The translation method lessens the visual burden compared with conventional methods without generating geometrical distortion. As in the parallel method, the optical axes of the cameras are parallel. What is different is that, based on the information of the right and left images, one of the two images is translated and moved so that the parallax of the target object is minimized. As a result, the locus of the target object is free from parallax and straight rather than circular. No geometrical distortion occurs and images with minimal parallax can be captured for the target object that the viewer wishes to focus upon.

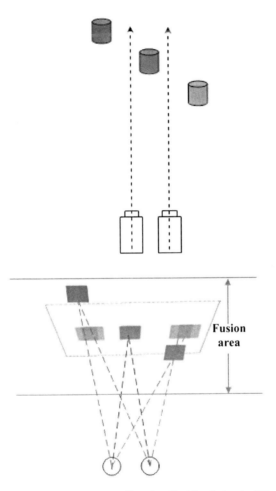

Fig. 7. Camera Layout(above) and Image Forming Position(below) of Translation Method

Principle of the Translation Method
We will demonstrate that the locus free from parallax is not circular but straight in our translation method.

[Demonstration]
Assumption: With the translation method, the right and left images are respectively translated and moved for the target point F in Fig.8 located at the center between the two cameras.

In Fig.8, point F is located at the center between the two cameras and on the straight line k which is parallel to the optical axes of the cameras. An arbitrary point A is located on line l which is drawn perpendicular to the straight line k. Images a1 and a2 of point A projected on the CCD planes lie on the line passing through point A and the imaging points (lens centers O1 and O2), respectively. In like manner, f1 and f2 are located respectively on the lines passing through point F and the lens centers. Since the two stereo cameras are of identical specifications, the distance between the CCD plane and lens center is equal for both cameras. Lines n and m connecting the two CCD planes and two lens centers, respectively, are parallel to line l where the objects are located. This means that ΔFAO1 comprising the line segment FA between the two objects and lens center O1 is similar to Δf1a1O1 comprising the line segment f1a1 between the two images on the CCD plane and lens center O1. Likewise, ΔFAO2 and Δf2a2O2 are similar. The similarity ratios for the two cases are identical. This means that the line sections f1a1 and f2a2 are of the same size or that point A is displayed equidistantly from center F on both images. Point F is displayed at the center between both images and free from parallax, and thus point A is also displayed in a parallax-free state. By using the translation method, the locus free from parallax is therefore straight rather than circular.

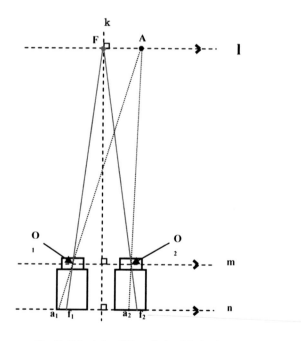

Fig. 8. Principle of Translation Method

As discussed earlier in the introduction, interdigital sensors can be employed in various applications depending upon the requirements. It could be used to measure the density of the material as shown in figure 5 (a), the distance between the material under test and sensor could be measured with the help of varying excitation fields as in figure 5 (b). It is also possible to identify the non-uniform or unevenly shaped materials using the interdigital sensors as shown in figure 5 (c) and they are also very good moisture sensors, as shown in figure 5 (d).

4 Experiments

Images obtained by the cross method and the translation method were compared to evaluate the effectiveness of the translation method and its capability to eliminate geometrical distortion in stereo images.

Geometrical distortion unique to the cross method varies with the convergence angle of the stereo cameras. The cross method and the translation method were used for near-distance and mid-distance image capture to learn which method was better at presenting stereo images for objects located at varying distances.

Outline of the Experiments

- Using the cross method and the translation method, images of objects at near-distance and mid-distance were captured to obtain a total of four types of images.
- Image capture environments were identical for both methods.
- The display range is narrow in the translation method. For this reason, the display range of the cross method was purposely narrowed so that when viewing with the naked eye the observers could not easily distinguish between the two methods.
- The observers were asked to evaluate on a five-level scale the stereoscopic effects, fatigue and ease of viewing the four types of images.
- The observers were asked to select which method appeared method for the respective distances.
- The images were reproduced for viewing in random order.

The results of the experiments are reported below.

1) Experiment 1: Near-Distance Image Capture

[Purpose of Experiment]
Evaluation of the effectiveness of the translation method for objects placed about 30cm in front of the stereo cameras.

[Description of Experiment]
Geometrical distortion increases when the cross point is set at a nearby object because of the large convergence angle. Distortion increases towards the edges of

an image. To fill the screen with the captured image of the object, we selected a bunch of flowers with a plenty of petals and leaves extending in all directions.

[Details of Experiment]

- Flowers were placed 30 cm in front of the cameras.
- For the respective image capture methods, the cameras were placed so that parallax was nearly zero at a point 30 cm in front of the cameras.
- The flowers were placed on a rotary table and turned 360 degrees in that position.
- The subjects were asked to watch the details of the object, including the tips of the leaves and flowers.

[Result and Observation]
The translation method received high evaluation marks compared with the cross method in all parameters of stereoscopic effects, fatigue and ease of viewing. In the overall evaluation, 75% of the subjects selected images captured by the translation method. These results indicate the possibility for eliminating geometric distortion by the translation method which would achieve a correct sense of distance compared with the cross method. We noted that the translation method was effective for enhancing stereoscopic effects, decreasing fatigue and improving the ease of viewing to some extent. The authors believe that this high evaluation comes from the absence of unpleasant sensations in the subconscious, which is the result of the lack of geometrical distortion in the space perception of the observers.

Table 2. Results of Questionnaire on Near-Distance Comparison

Near-distance	Translation method	Cross method
Stereoscopic effect	3.85	3.65
Fatigue	3.25	2.5
Ease of viewing	3.95	3.35
Overall evaluation	15	5

2) Experiment 2: Mid-Distance Image Capture

[Purpose of Experiment]
Evaluation of the effectiveness of the translation method for objects placed about 1 m in front of the stereo cameras.

[Description of Experiment]
Stereoscopic images with rapidly changing parallax have a pronounced 'jumping' effect, and are a favorite technique for producing content in entertainment media. The object was placed 1 m in front of the cameras and moved back and forth

quickly when capturing the images to see whether the translation method or the cross method was better for capturing images of mid-distance objects.

[Details of Experiment]

- The object used for image capture was a puppet.
- For the respective image capture methods, the cameras were placed so that parallax was nearly zero at a point 1 m in front of the cameras.
- The puppet was moved back and forth repeatedly in a distance range from 1 m to 30 cm from the cameras during image capture.
- To create differences in sensing the distance to the puppet, non-moving blocks were set at fixed positions in front of the puppet.

[Result and Observation]
Unlike the results of Experiment 1, the cross method was generally evaluated highly, although the difference compared with the translation method was small. The largest difference of evaluation marks between the translation method and the cross method was up to 0.4 marks in favor of the cross method for the fatigue parameter. This difference between the two methods in Experiment 2 was small, considering the fact that the difference was as large as about 0.7 marks in favor of the translation method for the fatigue parameter in Experiment 1. The cross method was relatively highly evaluated in Experiment 2 because of the reduction in the geometrical distortion, which is a problem inherent in the cross method. The convergence angle becomes smaller as the object moves away, and the geometrical distortion likewise decreases with decreasing convergence angle because the distortion depends on the locus of the horopter. This is the reason why the cross method was more highly evaluated in Experiment 2 than in Experiment 1.

Table 3. Results Of Questionnaire On Medium-Distance Comparison

Middle-distance	Translation method	Cross method
Stereoscopic effect	3.65	3.85
Fatigue	2.75	3.15
Ease of viewing	3.2	3.35
Overall evaluation	8	12

5 Conclusion

Our experiments have shown that the translation method is better than the cross method for capturing images of objects in the near distance. The cross method is slightly better than the translation method for capturing images of mid-distant objects. The translation method was originally intended for eliminating geometrical distortion when capturing images of near-distant objects, and as such, our experiments have successfully achieved their purpose.

The imaging field of the system used in our experiments was very narrow due to restrictions of the system's performance. Because geometrical distortion increases toward the edges of the screen, difference s in the two image capture methods can be more clearly demonstrated if we develop and perform the same experiments on a system capable of displaying larger images.

Based on the above experiments and discussions, the authors believe that their present study has demonstrated the possibility and significance of the translation method.

References

[1] Kim, H., Nishikawa, K., Takeno, J.: Research on Convergence Control and Reduction of Eye Fatigue. In: International Conference on Communication and Control Technologies CCCT 2004, The International Institute of Informatics and Systemics(IIS), Proceeding, vol. II, pp. 401–406 (2004) ISBN 980-6560-17-5
[2] Harashima, H.: Science of 3-dimensional pictures and human beings. Ohmsha Inc. Pub. (2000) (in Japanese)
[3] Grieco, A., Nolteni, G., Piccoli, B., Occhipinti, E. (eds.): The vertical horopter and the angle of view. Elsevier Science B.V., Amsterdam (1995)
[4] Ohkura, M., et al.: Measurement of Horopter and alleys in auditory space. In: Proceedings of the 2002 International Conference on Auditory Display, Kyoto, Japan (2002)
[5] Tachi, S.: Fusing Virtual Reality, Robotics and Networks Together. In: 14th International Symposium on Measurement and Control in Robotics
[6] Kawai, T., Shibata, T., Ohta, K., Yoshihara, Y., Inoue, T., Iwasaki, T.: Examination of a Stereoscopic 3D Display System Using a Correction Lens. Stereoscopic Displays and Applications 5006(47), 254–262 (2003)
[7] von Helmholtz, H.: Treatise on Physiological Optics. In: Southall, J.P.C. (ed.) vol. 3. Optical Society of America, New York (1925)
[8] Mizukami, M., Kim, H., Takeno, J.: Acquisition of Accurate Distance Information in Stereovision Using Move-parallel Method for Teleoperation. In: Proceedings of the IEEE International Conference on Mechatronics & Automation Niagara Falls, Canada, pp. 1197–1202 (2005)

From Labs to Real Environments: The Dark Side of WSNs

C. Alippi, C. Camplani, C. Galperti, and M. Roveri

Dipartimento di Elettronica e Informazione
Politecnico di Milano, Milano, Italy
{alippi,camplani,galperti,roveri}@elet.polimi

Abstract. Distributed environmental monitoring with wireless sensor networks (WSNs) is one of the most challenging research activities faced by the embedded system community in the last decade. Here, the need for pervasive, reliable and accurate monitoring systems has pushed the research to address aspects related to the realization of credible deployments able to survive in harsh environments for long time and not only toy applications working in laboratories. Designing an effective WSN requires a good piece of engineer work, not to mention the research contribution needed to provide a credible deployment. As a matter of fact, to solve our application, we are looking for a monitoring framework scalable, adaptive with respect to topological changes in the network, intelligent in its ability to react to evolutions in the external environment, power-aware in its middleware components and endowed with energy harvesting mechanisms to grant a long lifetime for the network. The paper addresses all main aspects related to the design of a WSN ranging from the possible- need of an ad-hoc embedded system, to sensing, local and remote transmission, data storage and visualization. Two applications, namely monitoring the marine environment and forecasting the collapse of rock faces in mountaineering areas will be the experimental leitmotiv of the chapter.

1 Introduction

In recent years, technological advances have allowed IT manufactory industries to design low power consumption and compact embedded systems able to collect and process information as well as communicate through a radio module. At the same time, advances in sensor technologies have provided devices easily interfaceable with microprocessors (when not directly integrated on-chip); thus, significant (from the application point of view) networks of intelligent sensors have been made possible. Such networks are composed of a high number of nodes, each provided with memory, sensors, processing capacity and a RF section for data exchange. Units automatically collect information even in environments that can be seen as "hostile" to human operators, or at least not easily reachable, and convoy data to a remote station for further processing. Such networks, known as *Wireless Sensor Network* (WSN), constitute a very interesting solution for many problems requiring a distributed monitoring, granting a level of detail- both spatial and temporal - non otherwise obtainable. The early studies propose the *Smart Dust* concept as an ideal model for WSNs [1] [2] [3]: thousands of tiny, possibly- air-dropped nodes [4] that "*can be moved by winds or can even remain suspended in air*" [5]. For a flashback to those days the interested reader can refer to [6] [7] [8].

Surely, a smart dust unit is the perfect instrument for a non invasive, low cost (and therefore pervasive), monitoring system. Unfortunately, this intriguing perspective is far from becoming reality and can, nowadays, only be intended as an asymptotical trend that, hopefully, the technology will reach. At the current state of the art the "cube-millimeter" [2] smart dust device is infeasible and represents a visionary commodity if energy harvesting, storage and antenna modules need to be integrated into the "cube". In parallel, the costs of the network, i.e., the cost of units, deployment and maintenance, increase quickly with the number of nodes and, for a large network(1000-10000 nodes [2]), can become unbearable [9]. However, it must be said that applications seen in the literature rarely show networks topology above 30 nodes; as such the topology is *very sparse* in contrast to the *dense* one we might expect.

Smart dust is thus an ideal, asymptotic model, yet provocative and appealing to the large public.

Designing an effective WSN requires a multidisciplinary team for the large variety of aspects implicitly addressed. People from the phenomenological world as well as measurements, electronics, embedded systems, telecommunications, software engineering, data bases and management just to name the few are generally needed to grant a successful not trivial- application.

Even if a node contains a simple processing unit, the technology as seen at the system level, is rather complex. As such, many improvements at the very local level in any of the research fields which can be identify might provide only an ϵ-improvement in the real context. We therefore suggest researchers to consider a system level approach, e.g., as implemented by the *cross-layer* optimization [10] [11] where the system design takes into account also the interactions among different layers of the system of systems (as a WSN should be intended).

Cross-layer optimization should then be considered also at the data level. There, layers correspond to the life cycle phases data undergo: acquisition, local processing, remote transmission, data storage, high level processing. A change in one of the envisaged phases, for example induced by the introduction of a new routing protocol, could impact on the interface towards local processing since different data transmission modalities might require a different behavior of the local node (e.g., time based, event driven).

These design aspects have been only partially addressed in applications presented in the literature, hence limiting the WSN technology credibility. We believe that a credible WSN is more than a prototypal application able to work satisfactory in a laboratory. Since it must live (even if for a short time) in a real deployment, it must assure adaptability to environmental and network topological changes, robustness w.r.t. faults (transient or permanent) and failures which may affect the network units, energy harvesting mechanisms for scavenging energy from the environment and effective storage means to accumulate energy.

At the same time WSN designers have to address both in-field and operational problems. The former related to the deployment (e.g., identification of the displacement of sensors, deployment difficulties, robustness and waterproofness of the case), the latter associated with the interaction of the monitoring system with the environment and, somehow, with the extension of the network lifetime (e.g., energy harvesting and management, faults and failures, thermal drifts, ageing effects).

In fact, the environment is an aspect not rarely underestimated in the literature: also a green hill can become an *harsh environment* under certain meteorological conditions. Rain, wind, sun, frost, snow and lightning can, in fact, damage partially or totally the networks nodes.

Here, we discuss the principal aspects associated with a distribute monitoring system and, in particular, those related to the design, implementation, best practice and deployment of a credible WSN-based system.

The structure of the paper is as follows. Section II provides a reasoned list of applications solved with WSNs, while Section III summarizes the methodological aspects associated with the design of a credible WSN. Finally, section IV presents two application cases in more detail: marine environment monitoring and rock collapse forecasting.

2 WSNs Based Applications: A Critical Survey

In the recent years, WSN have been considered to solve several applications ranging from environmental monitoring [12] [13] [14] [15] [16] [17], tracking [18], surveillance [19] [20] and fine monitoring of agricultural parameters [21]. The aim of this section is to critically discuss such applications by addressing their "credibility" in terms of adaptability to topological changes, availability of energy harvesting solutions and network lifetime.

[12] presents a star-based topology for seabirds habitat monitoring (the gateway collects data from the sensor nodes and forwards them to the remote control station). Here, for the first time, authors point out that energy harvesting is a fundamental aspect for a credible deployment: the gateway is augmented with a photovoltaic panel, leaving sensor nodes battery powered (resulting in an estimated lifetime of the network of six months). The communication (from sensing units to the gateway and then to the base station) is based on a simple half-duplex hierarchical protocol; the solution requires the gateway to be always on. No information about unit synchronization is provided.

A more complex WSN architecture is proposed in [13] envisaging a multi-hop WSNs for wildland fire monitoring. The limited adaptation ability of the monitoring system requires human intervention for introduction of new nodes, which are battery powered. No energy harvesting strategies are implemented.

In [14] a system for monitoring volcanic eruptions has been suggested; no energy harvesting solutions have been considered for the three units + gateway deployed network.

A wireless sensor network to monitor sub-glacier environments is presented in [15]. Sensor nodes, which are equipped with temperature, pressure and tilt sensors, are battery-powered and no energy harvesting solutions are considered impacting on the expected network lifetime which is estimated in few months.

The ARGO project [16] aims at monitoring temperature, salinity and see current profile to study the effect of climate changes on the oceans. Each sensor node is a free-drifting buoy equipped with temperature and salinity sensors and acquired data are transmitted to a remote control center by a satellite link. No energy harvesting solutions have been considered and, moreover, the cost of each node (15000 USD) is far from being cheap.

A simple WSN-based application is presented in [17] where a sensor network composed of 33 nodes was deployed on a single redwood tree. The aim of the research was to map the differences in the microclimate (temperature, humidity and solar radiation) associated with a single tree. The WSN units are battery equipped; neither energy harvesting systems nor robust routing algorithms are available.

[18] presents a WSN system for tracking the movements of Zebras through a GPS sensor taking samples every three minutes. Each WSN unit is equipped with batteries and photovoltaic cells. Unfortunately, the routing algorithm is very simple (flooding algorithm) and not effective in complex WSN applications.

Application [19] presents a WSN designed to detect the invasive and dangerous Cane Toad in northern Australia. The WSN units acquire data through acoustic sensors, locally process them with a FFT and transmit data to a gateway. In turn, the gateway sends only the presence/absence of the frog in the area. Neither energy harvesting or power aware solutions nor sophisticated energy management policies are reported.

[20] represents one of the major efforts in the WSN field. The application refers to the development of a WSN to identify mobile targets within a surveillance application. The 70 units also manage occurrence of failures by periodically rebuilding the routing paths. No energy harvesting solutions have been considered (but an effective duty cycle mechanism is envisaged).

[21] presents a WSN-based application to monitor the microclimate conditions of vineyards (e.g., temperature, soil, moisture, light, and humidity). Sensor nodes, which are deployed in a regular grid of about 20 m edge, are organized in a hierachical network aiming at conveying data to a remote control station. Sensor nodes are battery-powered and do not consider energy harvesting solutions. The expected network lifetime is less than a year.

3 Designing a WSN

As already discussed a real deployment requires to address relevant and numerous issues that do not emerge from simulations and only to a small extent arise in laboratory. These aspects can be grouped into, -not exhaustively- seven main categories: hardware, energy harvesting and storage, energy management, deployment, software, simulators and transmission.

3.1 Hardware

Since units must be as low invasive as possible their size should asymptotically tend to the smart dust concept. To obtain a reduced unit size we might require to create more than one electronic board, e.g., to host the energy harvesting mechanism and power distribution, the processing and radio module and the sensorial and signal conditioning one. For each of them particular attention must be devoted to the physical displacement within the board (e.g., a multi-layered board introducing additional cost) and of the board itself (e.g., piled up vertically) and how to connect the buses among them (presence of wind and vibrations might even require to consider particular connectors to prevent the electronic contacts from loosing). Another important issue is the choice of electronic components within the *commercial, industrial* and *military* standards. Units must always guarantee the correct functioning even in harsh environments. Commercial electronics generally operate in the 0C, 60C range but, if our unit is inserted in a metallic case exposed to the solar radiation, then we could easily encounter an unwished error induced by overheating. In such a case we should opt at least for components satisfying the industrial standard.

Electronics should be environmentally friendly and not contain polluting substances (eventually units might be disposed and wasted in the environment). In this direction the "Restriction of Hazardous Substances Directive" (RoHS) [22] limits the use of hazardous materials in electronic and electrical equipments. Off limits elements are lead, mercury, cadmium, hexavalent chromium, polybrominated biphenyls and polybrominated diphenyl ether where the maximum permitted concentration are below 0.1% (or 1000 ppm) - except for cadmium, which is limited to 0.01% (or 100 ppm). We surely suggest the designer to use RoHS compliant electronic components (even if your legislation does not require RoHS compatibility).

A last note about sensors. Sensors to be used in a WSN unit obviously depend on the application needs. The final one is defined at design time based on cost, accuracy and complexity (a piezoelectric accelerometer is more accurate than a MEMS-based one but its cost is significantly higher). The designer should balance accuracy with cost, always keeping in mind that accuracy should be matched with that required by the application (do we really need high accuracy when the decisional process is affected by a significant uncertainty?).

In addition to application-specific sensors, we suggest to include inside the unit some utility sensors (e.g., a humidity and a temperature sensor) to monitor possible critical situations (e.g., the infiltration of water and/or moisture or excessive temperature). Moreover, we found particularly useful the use of a magnetic switch for enabling the unit and keeping the external package sealed. The unit package (or case) is one of the most important and underestimated (at least in the academia) issue in the design of units able to survive outside. Here, the main aspect is watertightness; as such, special connectors, screws, radio antenna slots and fixing solutions must be envisaged to prevent the percolation of humidity and water. In case of environments characterized by wind or waves we suggest the use of flexible antennas (and, possibly, a rubber joint) being less prone to breakings.

3.2 Energy Harvesting and Storage

The credibility of a wireless sensor network deployment is somehow related to the energy supply which, generally provided by batteries, is a finite resource. Even if wireless sensor nodes are mostly planned for long term outdoor operations, only seldom rely on external energy supply sources (which, in turn, require the node to consider energy harvesting mechanisms). In principle, all energy sources should be exploited to extract the available energy; among the others [23] [24] [25] [26] [27] the solar one is generally the most effective in outdoor applications for the high power density provided and exploitable through photovoltaic cells. Despite the fact that solar energy provides much more power compared to alternative sources (e.g., vibrational and acoustic) very few attempts have been made to extract energy from solar radiation in wireless units. Up to date, the maximum power density obtainable from a modern solar cell is about $5 - 20 mW/cm^2$ (outdoor, sun at the zenith) whereas all other sources provide an energy gain far below $1 mW/cm^3$ [28]. Unfortunately, solar cells exhibit a strong non-linear electrical characteristic, which makes it low effective to extract energy in non optimal radiating environments. Energy availability can be associated with weather condition changes (e.g., cloudy and not optimally radiating solar power environments), aging effects or efficiency degradation in the solar panel (e.g., dust or rust on the cell surface). Moreover, the energy transfer mechanism is strongly influenced by the illumination condition such as the angle of incidence of the sunlight which varies along the day. All such phenomena can be interpreted as transient or permanent perturbations affecting -and reducing- the efficiency of the energy production phase. To overcome these problems and optimally exploit the energy generated by a solar cell, effective solutions must be considered, e.g., the ones based on maximum power point tracker (MPPT) circuits (e.g., [29]). MPPT circuits are special power circuits able to convey energy generated inside a solar cell into a storage element, like a battery, while aiming at maintaining the working point of the cell around the optimal one (maximizing the transferred power).

The energy generated from the solar cell is then transferred to an energy storage device, acting as energy buffer, and made available to the WSN unit.

Nowadays, the most common electric energy storage device is represented by chemical batteries which rely on electrochemical reactions to store energy in chemical bounds. They are cheap but the charge/discharge curve is non-linear, time-variant and batteries require full discharge/recharge cycles to maximise their lifetime.

An unavoidable problem in an on-the-field deployed embedded device is that electric energy is generated at a not controllable power rate leading to severe deviations from battery manufacturer's specifications dictating the charging current profile. Furthermore, we have the partial charge/discharge effect, here suffered by batteries during the day/night solar radiation cycle [30]: during the day solar power is fed into batteries and recharges them while during the night no solar power is available and batteries are discharged by the system power consumption. This phenomenon is amplified in the case of not optimal radiation due to bad weather conditions.

In order to mitigate the partial charge/discharge phenomenon and the charging current profile problem, two technological solutions have been recently proposed in the literature [31]: tandem batteries and supercapacitors.

The basic idea behind a tandem battery solution [31] is to separate in time the charge and the discharge phases, i.e., letting the batteries be charged and discharged in separate and definite intervals of time. As a consequence, while one battery pack powers the system (and thus is obviously discharged) the other battery pack is recharged by solar power; when appropriate conditions occur battery packs' operational rules are inverted.

Supercapacitors, namely capacitors with a very high capacitance (up to 3000F), represent the new generation of energy storage means. Being capacitors, the residual energy can be easily evaluated through the voltage across the capacitors itself whereas for traditional batteries can be roughly estimated with an accuracy decreasing with the battery aging. Moreover, batteries have a long charge/energy extraction constant of time supercapacitors have not. This is a first enormous benefit of using supercapacitors in storing energy instead of batteries since they don't require a particular current profile: once we have energy we can simply store it at any rate. At the same time, when energy is required, we can simply extract it at any power level.

A second important advantage of supercapacitors over batteries is their lifetime. Since supercaps are not based on chemical reactions but solely on electrostatic field generation, they are able to withstand from hundreds of thousands up to millions complete charge/discharge cycles. Batteries, due to a number of secondary chemical effects, seldom withstand more than one thousand complete charge/discharge cycles.

3.3 Energy Management

Since energy is a finite resource, a careful energy-aware design of hardware and software components is requested. In general, the wireless communication mechanism is a prime energy eager aspect (in addition to flash memory storage). As a consequence, several energy management techniques have been proposed aiming at minimizing the radio activity (a detailed survey can be found in [32]). Different solutions envisage data compression [33] [34] and aggregation [35] [36], predictive monitoring [37], topology management [38] [39], adaptive duty cycle [40] just to name the few. Differently from what believed, energy consumed by sensors and processing units is not necessarily negligible compared to the one needed for communication. On the sensor side this happens for three main reasons [41]: 1) highly energy consuming A/D converters (e.g., acoustic or seismic sensors require high-rate and high-resolution A/D converters); 2) sensing arrays (e.g., CCD or CMOS image sensor); 3) active transducers (e.g., sonar and radar). Several solutions for energy management at the sensor level, e.g., see [41], have been suggested in the literature. For instance, multi-scale sensing [42] [43] relies on sensors characterized by different resolutions. Accuracy can be traded off with energy efficiency by using a low-power sensor to acquire data; model-based active sampling [44] takes a model of the phenomenon to be monitored so that sensor nodes may decide whether to acquire or to estimate a new sample (thus saving the energy for the new sample acquisition); finally, the adaptive sampling approach [45] [46] [47] [48] aims at adapting the sampling rate to the current temporal dynamics of the phenomenon to be monitored directly during acquisition time. Of course, an energy management at sensor level is not the panacea for a credible deployment of WSNs; what suggested here must be integrated with available energy management techniques acting at various abstraction levels (e.g., [?] proposes a *power state machine* which allows to define effective energy policies by modeling the energy consumption of nodes in terms of states and transitions; [?] proposes the dynamic voltage scaling of the processor in embedded systems to save energy). The interested reader can refer to [49] for a detailed review.

Recently, a novel trend within the WSNs community sees samples more than a datastream to be conveyed to the control room. In fact, once suitably extracted from data and processed, features can be used to improve the efficacy and efficiency of the network (e.g., they constitute the basic elements to generate hierarchical event triggering, and distributed decision making). This feature-centric approach is well justified by observing that most of physical phenomena satisfy the locality properties: temporal locality (samples are related in time) and spatial locality (close views of the phenomenon provide related data). In this direction only few results have been published in the WSNs arena. For instance, [50] suggests a theoretical framework exploiting spatial and temporal correlations in data to be used for developing effective routing algorithms. Among other hypotheses, the physical phenomenon is assumed to be ruled by a Gaussian distribution with known variance. A correlation-based compression technique for routing algorithms is proposed in [51] which relies on an empirical approximation of the joint entropy between nodes w.r.t. their distance. [52] shows a Medium Access Control layer which exploits the spatial correlation framework of [50] to reduce the number of transmitted data. A different approach is proposed in [53], which suggests exploiting the correlation among subsequent frames in wireless multimedia sensor networks to reduce the amount of data to be transmitted and stored. The ability to measure the "mutual affinity" among spatial measurements [54], i.e., cross-dependency or correlation, would allow researchers for defining a new generation of solutions to routing, decision, and energy management. In fact, information-based routing algorithms could generate clusters by grouping units according to the mutual affinity of features. Instead, decision making algorithms could exploit the mutual affinity within a cluster to make decisions, e.g., detecting the presence of an event by relying on cluster-based features. Finally, an effective energy management policy for units would exploit mutual affinity to switch off those units or sensors whose information-content is provided indirectly by others.

3.4 Deployment

The deployment of the sensor nodes should be addressed in strict conjunction with the experts of the physical phenomenon under monitoring. In fact, an ad-hoc deployment (where the positions of the sensor nodes is defined from the phenomenological point of view) could provide a more meaningful information content about the phenomenon than a random deployment (where sensor nodes are deployed in random positions).

Moreover, even if the deployment phase is strictly related to the specific application, two general aspects can be identified.

At first, the deployment of units must be adequately signaled to prevent incidents (e.g., in aquatic environments buoys must be equipped with a proper signaling).

Second, the environment in which WSN units are deployed may change over time and affect the units (e.g., terrain landslide, presence of growing vegetation). In fact, dust, encrustations and sand transported by the wind can cover the WSN units, mould and algae cover the sensors, thermal stress can change the system performance.

Even if adaptive solutions can be envisaged to reduce the impact of the interaction between monitoring units and the environment a loss in performance must be expected. The designer should try, wherever possible, to guarantee a graceful degradation of the network performance (e.g., with a MPPT-based solar harvesting, adaptive sensor calibration, etc.).

3.5 Software

Designing and developing reliable and reusable software is a main issue for WSNs. Since the technology is rather new it lacks in SW tools for easing the application development and testing; not rarely, oscilloscopes and frequencies analyzers are needed to verify the correct functionalities at the lowest level of the application code both at the microprocessor, low power DSP (where present) and radio.

Still open issues from the software point of view can be summed up as: the specification of programming patterns for the embedded software, the definition of a framework for a rapid prototyping and debugging of WSN-based applications and a multi-layered testing methodology.

Nowadays, the programming approach relies on an "*embedded*" philosophy to design and implement the system software, where each aspect of the software is tailored to the specific context, e.g., acquisition, routing, storage etc. The main advantage of this approach is a shorter developing time compared to that needed when a more general approach is considered; unfortunately, developed parts are hardly reusable in different application contexts.

Even if more articulated engineered approaches exist in the literature, e.g., see [55] [56] [57] [58] a complete coverage of all aspects involved in a WSN-based application still remains an open challenge for the research community. In this direction, a middleware supporting such features would be a breakthrough in software development for WSNs. Table 1 shows the main differences among the above cited middleware for WSNs. It emerges that DSN [57] and TinyDB [55] provide a full support to low level software whereas GSN [56] and SWORD [58] offer only an high level interface for devices (TCP/IP + xml-based protocols). Conversely, only GSN and SWORD support the heterogeneity by allowing hosting and managing different devices. However, such flexibility is "static" in the sense that the system needs to be configured at deployment time to use different technologies. If the user needs to insert a new type of node after the deployment, the system has to be stopped and reconfigured.

On the contrary, PerLa [59], a novel middleware for WSN, provides a database abstraction of the entire system. Each device can be queried through PerLa with its simple SQL-like interface, where data and commands can be read/sent to the device without requiring specific information about the device. PerLa wraps completely the device communication protocol that becomes transparent to high levels. Moreover, PerLa addresses the problem of heterogeneity at run time by using a xml meta-description language for new node types. This description is parsed at run-time by the middleware that builds and loads a wrapper to handle the communication with the device. After the binding with PerLa, the device is ready to be queried. A first use of PerLa in a real environment is presented in Section 4.2.

Another limitation WSN-designers have to face is the reduced availability of software development and debugging tools, which are limited in many cases to the C compiler and the oscilloscope. The software running on nodes is generally, specific to the application and platform dependent. Moreover, due the limitation in terms of computational and memory capacity, the software must be "light" and optimized for the envisaged hardware platform. Starting from the scratch, an implementation of a software for embedded distributed systems can become quickly a nightmare. However, several frameworks have been proposed to address such a problem. In particular TinyOS, despite its limitations [60], is widely used and many middleware [55] [57] are based on it. The most important

Table 1. Comparison table of middleware for WSNs

	TinyDB	GSN	DSN	SWORD
Data acquisition	√	X	√	X
Configurability	--	X	X	X
Data aggregation	√	--	√	X
High level integration	√	√	√	√
Re-Usability	--	√	--	√
Low Level software support	√	X	√	X
Deployment-time heterogeneity support	X	√	X	√

feature of TinyOS is the design of a hardware abstraction layer model (HAL) that allows, in many cases, the 100% portability of the code among the supported platform (i.e., MicaZ, Telosb, etc..). However, the entire system is written in a special language, the nesC, which is not trivial and error-prone due to its event-driven paradigm. Moreover, kernel and user spaces are not separated (e.g., the programmer has visibility access to the scheduler), occurrence that increases the possibility of generating *hardcore*-bugs.

Finally, the test phase is generally performed by testing one system layer at time by using mock objects to simulate inlayer interfaces. However, this approach does not solve the problem that all modules of the system should be tested together. For sure, module integration-bug is the most nasty thing to manage, and we experienced that not rarely it requires the unorthodox use of oscilloscopes and digital analyzers. It is obvious that a general purpose testing and debugging technique would be very much appreciated. Moreover, even if some network simulator (see next Section) provides tools to simulate the software running on nodes, driver or hardware related bugs cannot be identified by simulation and must be discovered only on real nodes.

3.6 Simulators

The need to improve the design and speed up the deployment phase requires the development of simulators for WSN aiming at evaluating, at design time, the performance of the final system. As stated in [61], this goal seems quite straightforward but, in the practice, is not easily achievable due to two main reasons. At first, the simulation overhead is not negligible, hence limiting the scalability of the WSN simulator in terms of number of units. Secondly, accurate results require a fine modelling of all aspects (e.g., the hardware platform, the radio module, the sensors, the routing protocol, the physical phenomena under monitoring, the environment in which the WSN nodes operate) that may affect the performance of the wireless sensor network.

The simulators can be grouped into two main approaches: general purpose and specific simulation environments.

The one following the first approach represent general communication network simulators which have been extended to manage WSNs.

NS-2 [62] is a discrete-event general purpose network simulator which uses OTcl as control and configure scripting language. Unfortunately, it does not scale very well w.r.t. the number of nodes, thus providing poor performance for large ad hoc wireless networks [63].

J-Sim [64] is a Java-based general purpose, opensource simulation framework initially developed for wired networks and then integrated with a specific WSN package. Unfortunately, it supports only the 802.11 MAC layer and some high-level models for the processing unit, radio model, battery and sensors.

OMNeT++ [65] is a discrete-event, general purpose simulation environment with an intuitive GUI. The Mobility Framework package allows OMNeT++ to simulate ad-hoc networks. Unfortunately, the connection between nodes of the network should be a-priori fixed by designers and cannot be dynamically modified. On the contrary, specific simulators target results to a specific hardware and operating system. This has two main effects: a) the simulated code also run on real nodes (or only minor modifications are needed) and, b), the simulation provides more accurate results.

In this second class we find TOSSIM [66], which is a simulation enviromente able to emulate NesC code for TinyOS applications running on MICA Motes. TOSSIM guarantees a good scalability w.r.t. the number of nodes but to the detriment of the heterogeneity of the network. In fact, it is not possible to differentiate the behavior of the nodes in the network since they must share the same code.

ATEMU [67] emulates codes running on AVR motes. It guarantees more detailed and accurate results than TOSSIM at the expenses of scalability (few nodes can be simulated).

EmStar [68] is a Linux-based simulation framework able to simulate both MICA2 Motes and iPAQ-based microservers and, differently from other simulation environments, provides interfaces to connect the simulation with existing hardware. Unfortunately, it relies on a very simple environmental model and network medium.

3.7 Transmission

Quality of service of communication among units is affected both by meteorological and environmental conditions as well as electromagnetic pollution. Rain, clouds or humidity (but also trees and vegetation) influence negatively the communication. In addition, the presence of metals in the neighborhood of antennas (e.g., wire net, metal posts) causes the formation of multi-path that might heavily corrupt the data transmissions. Designers have also to take into account the presence of wireless networks that operate at the same frequencies of the WSN units (e.g., some WiFi networks work at 2.4GHz) and can severely disturb communications. For such reasons and by also considering that WSN units generally operate in a harsh, dynamic environment requiring the network to change over time to face permanent or transient node faults, failures and environmental changes, adaptive power-aware multi-hop routing algorithms [69] must be considered to provide topology adaption in an energy-aware context. These algorithms are also suitable to manage the unpredictable effects of energy harvesting mechanisms where units can run out of power with a subsequent unwished disconnection from the network; for some others the opposite holds, in the sense that units, back with energy, need to be reconnected. Moreover, the use of multi-hop routing algorithms [70] [71] can increase the monitoring area. In fact, in single-hop routing algorithms, units transmit directly to the base station (this limits the deployment area to the maximum transmission distance of the radio module). On the contrary, multi-hop routing algorithms, which allow the presence of intermediate routing units, guarantee an easy scalability of the WSN network at the expenses of an increased complexity of the routing algorithm.

Table 2. Suggestions for the design of a credible WSN-based applications

Issue	Suggestion
Hardware	• Consider the package sizes in the design of the electronic boards • Use industrial electronic components and boards able to operate in the application range • Use RoHS certified electronic and electrical equipments • Test the electronic boards • Sensor calibration • Design watertight packages • Consider in the design possible loss of acid from the batteries • The batteries warm up during the usage. Keep them separated from the electronic boards
Energy harvesting	• Energy scavenge must be pursuit to grant long time operation • Use MPPT circuits for optimally transferring solar energy to storage means in adaptive radiating environments
Deployment	• Signal adequately the WSN units in the environment • In case of marine environments, use elastic mooring cables and specific hooks to maintain horizontal the buoy. Moreover, set the length of the mooring cable to the high tide but distance adequately the buoys to prevent twists
Transmission	• Consider the possibility that wireless network (e.g., WiFi) can corrupt the transmission • Keep antennas far from metallic objects • Use adaptive and power-aware routing algorithms • Use standard transmission protocols
Software	• Design reusable code as much as possible • Take into account in the software the time required by electronic devices to wake up • Foresee unorthodox software debugging methods (e.g., use the oscilloscope to verify the correct switching on/off of electronic devices)

4 Two Applications: Monitoring the Marine Environment and Forecasting a Rock Collapse

We present two WSN-based monitoring applications whose designs required addressing much of the issues critically discussed in the previous sections: the Australian coral reef monitoring and the rock collapse forecasting on the alps.

4.1 Coral Reef Monitoring

The great barrier reef, one of the most precious and sensitive environmental systems, has been classified by UNESCO as a world natural masterpiece. However, researches believe that the human impact and the climate change is affecting the health status of the coral reef up to a point that, without any major important change, its expected lifetime should not exceed 40 years. Effective intervention actions require first a deep quantitative study of the marine ecosystem through phenomenological measurements, accurate models and analysis of the close interactions between climatic evolution and environmental impact. The coral reef harsh environment, its spatial area coverage and existing limited (semi)automatic technologies require a continuous human intervention in the monitoring phase which, due to cost and difficulties, have contained the phenomenological study. To address this issue we developed a WSN [72]. In particular, the prototype application aims at monitoring temperature and brightness in the water at different depths in a circular $2800m^2$ sea area in Moreton Bay (Brisbane) see Figure 1.

In its simplest architecture the WSN is characterized by a star topology with sensor nodes sending their measurements directly to the gateway (single-hop transmission). The developed WSN is composed of 9 sensor nodes that are inserted in buoys anchored to the sea basement. The base station has been positioned at the Moreton Bay Research Station and Study Centre of the University of Queensland in North Stradbroke Island Brisbane. Both gateway and sensor nodes endow energy harvesting mechanisms with MPPT. The distance between gateway and base station is approximately 1Km. In marine applications, the sea water is a trouble element for radio communications. The creation of superficial electrical currents during communication severely decreases the signal-to-ratio of the transmitted signals. Moreover, the antenna oscillates with the buoy causing an unstable line-of-sight between transmitter and receiver. We opted for both isotropic antennas and a robust transmission protocol to reduce the effects of buoys movements on the transmission. A modified TDMA method which includes power-aware and network scalability issues has thus been designed and developed. In particular, the suggested power-aware TDMA provides an efficient radio management both of the sensor nodes and the gateway (the gateway only listens the messages of the sensor nodes connected to the network). A robust registration phase has been included in the developed protocol to allow an efficient insertion/removal of sensor nodes to the network. The WSN deployment in the Coral reef has been particularly critical due to the anchorage system, the high/low tide and the involved materials.

Fig. 1. Final deployment at Moreton Bay: Two units and the gateway

First, buoys are generally anchored with a mooring cable to a reinforced concrete block displaced at the bottom. Unfortunately, the strength of waves may induce the buoys to flutter hence affecting the mooring cable (causing the loss of the buoys) or the deformation of the gasket of the package (causing the possible seepage of water inside the package). Moreover, when the distance between adjacent buoys is not adequate, the mooring cables may get twisted causing possible tears. Second, the high/low tide heavily affects length of the mooring cable. If the cable is too short, in case of high tide, buoys may remain under the level of the water, increasing heavily the possibility of seepages. On the contrary, in case of low tide, if cables are too long, they can get twisted increasing the possibility of cable breaks. We set the length of the mooring cable to allow the flotation even with high tide and outdistanced the buoys to avoid possible cable twists. Third, the formation of seaweeds on all the underwater elements (e.g., mooring cables, part of the buoys under the waterline) might weaken them (and making them more fragile to knocks and blows). The technological infrastructure for the processing and the transmission unit in the proposed framework relied on the Crossbow MICAz [73]. This sensor node provides a 7,37 MHz ATMega 128L processing unit (with 4KByte for the RAM and 138 KByte for the program memory) and a C2420 transceiver radio (single-chip 2.4 GHz IEEE 802.15.4 compliant RF transceiver with a 250 kbps effective data rate). The technological infrastructure we rely on for the radio link is the MaxStream 2.4GHz XStream Radio Modem [74]. Thus, the gateway uses the CC2420 radio to communicate with the sensor nodes (local transmission) and the XStream Radio Modem to transmit remotely the data - the measurements received from the sensor nodes together with the gateway measurements - to the control center (remote transmission). The sampling frequency of both the temperature and the brightness sensor is 1Hz; the acquired measurements are then averaged (to reduce acquisition noise) and sent to the gateway every 30s. Each DATA message is composed of 24 bytes with 14 bytes of payload. Figure 2 and Figure 3 show the measurements acquired by a node in four days acquisition campaign (from 2007/11/18 20:06:03 to 2007/11/22 09:20:27). In particular, Figure 2 presents the temperature, the brightness and the solar power generated by the PV cells, while Figure 3 shows the state of the batteries.

The brightness and the solar power in Figure 2 clearly shows the day/night period over acquisition days (e.g., we see the night between sample 0 and sample 960, the dawn and early morning between sample 960 and sample 1260 and the afternoon between samples 1800 and 2100). The temperature follows well the day/night periodicity but with a delay due to the thermal inertia of the water. Figure 3 shows the state of the batteries and the solar power in the acquisition period. At the beginning of the experiment Battery 1 is active; its voltage decreases w.r.t. time up to sample 900. Battery 2, at the same time, suffers from self-discharge phenomena causing a reduction of the voltage even when the battery is not used.

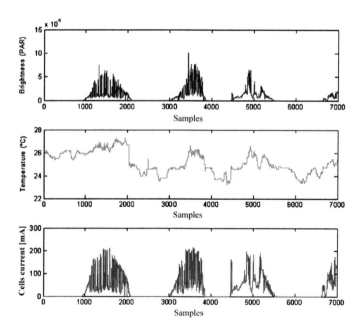

Fig. 2. Node 1: Brightness, temperature and solar power w.r.t. time (7000 samples between 2007-11-18 20:06:03 and 2007-11-22 09:20:27)

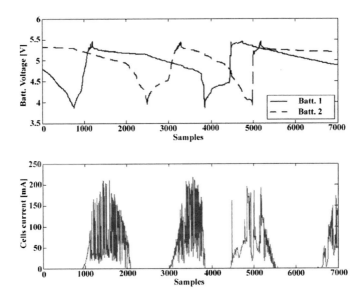

Fig. 3. Node 1: Voltage batteries and solar power w.r.t. time (7000 samples between 2007-11-18 20:06:03 and 2007-11-22 09:20:27)

When the voltage of Battery 1 decreases below 4V (sample 900) the energy harvesting mechanisms module switches between the two batteries and activates Battery 2. Between sample 900 and 1200 Battery 1 is recharged by the solar energy. When the voltage of the recharged battery overcomes 5.5V, the battery is cut off from the energy harvesting mechanisms module to prevent overcharge phenomena (sample 1200). As a consequence, Battery 1 is neither used (being active Battery 2) nor under charge and suffers from self-discharge phenomena (samples in the 1200 and 2600 interval). Battery 2 is used up to sample 2600 when the voltage decreases below 4V. Here, again, batteries are switched: Battery 1 is used and Battery 2 is under charge. The new sensor system has lead to the following direct outcomes: 1.ability to monitor the environment at multiple scales simultaneously so that it is both possible to validate existing models as well as define the future of management focussed modelling for these ecosystems; 2.a significant improvement in the prediction of the occurrence of ecological phenomena such as toxic algal blooms and climate-related loss of species and overall biodiversity; 3.ability to produce integrated multi-scale models and understandings of physico-chemical conditions (for use in underpinning both research and management of coastal ecosystems); 4.availability of precious and timeliness data for the development of risk-based and early warning systems (e.g., hurricanes formation) to underpin management response and the longer-term sustainable management of coastal marine resources.

4.2 Rock Collapse Forecasting

Monitoring civil buildings, unstable cliffs, slopes and rock faces is crucial to evaluate the dynamics of structures as well as provide data for stability and structural properties assessment (for a possible infrastructure collapse forecast). These aspects have been addressed in the related literature with important monitoring techniques mainly based on wired solutions (e.g., see [75] [76]). However, the need of wireless solutions in several applications is pushing research towards wireless sensor networks (despite the fact that, up to now, several still-open issues need to be addressed to make the technology credible, e.g., clock synchronization, effective energy harvesting mechanisms, power-aware node and network management, real time execution).

The WSN-based monitoring application aims at providing an ICT infrastructure for a real-time forecast of rock collapse in the alps. The deployment of the WSN is foreseen for July 2009 and the area to be monitored is part of the San Martino mountain which insists over the town of Lecco (Italy) - see Figure 4 for the area under monitoring. Collapse of rock faces is related to the geological substrate (or the nature of the concrete); the presence of fractures influences the stability of the envisaged material and their enlargement represents the first macroscopical step towards the collapse. The crack formation and enlargement is induced by a local micro acoustic/seismic activity, a phenomenon associated

Fig. 4. The area of deployment at the San Martino mountain, Lecco (Italy)

with the hierarchical organization of fractures inside the rock [77] or concrete-based structures. Micro acoustic bursts are generated in correspondence with the fracture evolution and cover a bandwidth up to 10kHz [78]. An example of a micro acoustic burst is presented in Figure 5.

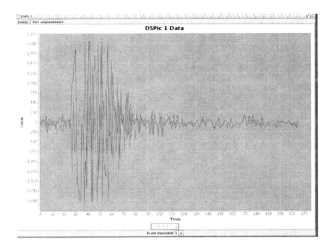

Fig. 5. An example of micro acoustic burst

Microacoustic signals can be acquired through geophones, piezoelectric or MEMS accelerometers (all solutions are feasible in a wireless sensor network use) differentiating in accuracy and cost. More in detail, geophones are surely the most sensitive sensors but, relying on a seismic mass displacement reading mechanism, are of large dimensions; on the other hand, they are passive, with energy consumption associated solely with the signal conditioning and the ADC conversion. Piezoelectric sensors are sensitive, provide a good noise immunity and a wide bandwidth (more than 10kHz, depending on the particular sensor) but their cost is high. Due to the different peculiarities both geophones and accelerometers are present in the developed WSN application. In addition, WSN units may be equipped with inclinometers, crack extensometers and temperature sensors to provide a comprehensive observation of the whole geological phenomenon. The design of the WSN-based application has been a real challenge: single and multi hop WSNs have been considered and a combined wired and wireless solution based on a field bus has been proposed. The architecture of the designed WSN is depicted in Figure 6. The Sensorial and Processing Units (SPUs), which aim at acquiring the measurements about the micro acoustic bursts, are grouped into clusters. All SPUs of a cluster area are connected with an industrial bus (i.e., the CAN bus) to a Bus Controller present on the WSN unit (BC-WSN) and use the bus to locally transmit the acquired data. Each BC-WSN then forwards the data received from the connected SPUs to a $2.45GHz$ ZigBee Gateway (GW).

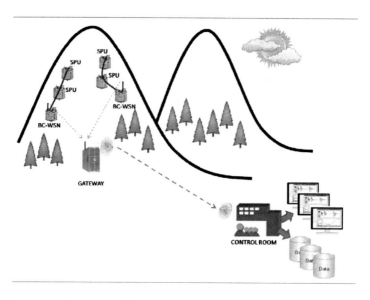

Fig. 6. Architecture of the rock collapse forecasting WSN application

Both the SPUs and the BC-WSNs are deployed on the rock face, while the Gateway is 100-400m far apart the WSN and deployed in a rockfall safe area. The Gateway then collects data coming from the BC-WSNs and forwards them to the Control Room (located at Campus Point laboratory - Politecnico di Milano - Lecco, Italy) with a 5GHz radio link. The distance between Gateway and Control Room is approximately 2.5Km. In a simplified design BC-WSN can directly act as GW and communicate directly to the Control room (first deployment).

The design of the WSNs unit was severely influenced by the critical deployment area since both SPUs and BC-WSNs are displaced on the rock face. The units have been designed only with RoHS compliant and industrial electronics to tolerate a very high temperature range. At the same time, to cope with strong winds, they are fixed into the rock face through a steel nail. Mounting of the units required the local Alpine group (the "Ragni" of Lecco). Particular attention was posed to the presence of lightning phenomena, which could completely destroy the deployment; in this direction a lighting protection system is being designed.

Identification of the displacement area required a-priori knowledge about the phenomenon under study as well as identification of the displacement areas for units to maximize the information content coming from sensors. Figure 4 presents the monitored area and the position of sensors as defined by geophysics researchers. Geophones, MEMS and strain gauges are distributed along the preexistent fracture.

Since MEMS sensors are positioned inside the case for protection, the case itself has been designed to avoid mechanical resonance in the sensor bandwidth (see Figure 7); a vibrational analysis has been carried out to identify the transfer functions of the case and the most appropriate glue was identified to fix units in the drilled 10mm ∅ hole.

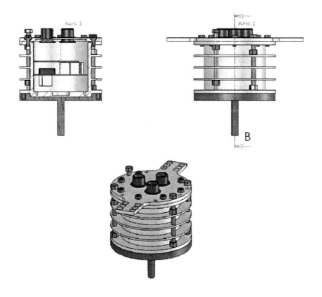

Fig. 7. Case

The SPUs software is mainly composed by acquisition, processing and communication layers. In the acquisition phase the SPU reads the ADC values, which are digitally filtered, while a trigger based on the energy content identifies the presence of bursts. Bursts are recorded into a memory buffer and transmitted to the BC-WSN unit (a μLinux-based board). The communication stack provides the basic functions allowing the CANbus master to retrieve data and send parameters and command from/to SPUs (the communication stack is cross platform).

The energy management of the WSN operates as follows: the transmission between SPUs and the BC-WSNs on the CAN bus is performed only in case of micro-acoustic burst detection, otherwise it remains switched off to save energy.

The GW communicates with the Control Room where is installed the middleware PerLa (presented in Section 3.5) to transform high level commands to a suitable form for the device and collects and interprets data coming from nodes. At the Control Room, data are then processed [78] to detect the presence of micro acoustic bursts and, in case of rock collapse danger, an alarm is forwarded to the organizations with jurisdiction (e.g., Fire Department, Civil Protection).

5 Conclusions

Design of credible WSN-based application requires to address challenging and multidisciplinary issues that rarely emerge with the simulation (or the testing in laboratory) of the WSN. The aim of the paper is to critically discuss the sometimes hidden and forgotten- aspects of WSN-based applications. The critical discussion is further integrated with a set of suggestions that, in the opinion of the authors, could be a valuable guide for designers to develop of credible WSNs. Two real WSN-based applications (monitoring the marine environment and forecasting the collapse of rock faces in mountaineering areas) have been presented as case study.

References

1. Kahn, J., Katz, R., Pister, K.: 'Next century challenges: mobile networking for "Smart Dust". In: Proceedings of the 5th annual ACM/IEEE international conference on Mobile computing and networking, pp. 271–278. ACM, New York (1999)
2. Warneke, B., Last, M., Liebowitz, B., Pister, K.: Smart dust: communicating with a cubic-millimeter computer. Computer 34(1), 44–51 (2001)
3. Kahn, J., Katz, R., Pister, K.: Mobile networking for smart dust. In: ACM/IEEE Intl. Conf. on Mobile Computing and Networking (MobiCom 1999), Seattle, WA, pp. 271–278 (1999)
4. Anderson, R., Chan, H., Perrig, A.: Key infection: Smart trust for smart dust. In: Proceedings of the 12th IEEE International Conference on Network Protocols. ICNP 2004, pp. 206–215 (2004)
5. Romer, K.: Tracking real-world phenomena with smart dust. In: Karl, H., Wolisz, A., Willig, A. (eds.) EWSN 2004. LNCS, vol. 2920, pp. 28–43. Springer, Heidelberg (2004)
6. Estrin, D., et al.: Instrumenting the world with wireless sensor networks. In: Proc. of the IEEE Acoustic, Speech, and Signal Proc. Conf., vol. 4, pp. 2033–2036 (2001)
7. Akyildiz, I., Su, W., Sankarasubramaniam, Y., Cayirci, E.: Wireless sensor networks: a survey. Computer networks 38(4), 393–422 (2002)
8. Sohrabi, K., Gao, J., Ailawadhi, V., Pottie, G.: Protocols for self-organization of a wireless sensor network. IEEE Personal Communications 7(5), 16–27 (2000)
9. Gamage, C., Bicakci, K., Crispo, B., Tanenbaum, A.: Security for the Mythical Air-dropped Sensor Network. In: Proceedings of the 11th IEEE Symposium on Computers and Communications, pp. 41–47. IEEE Computer Society, Washington (2006)
10. Van Hoesel, L., Nieberg, T., Wu, J., Havinga, P.: Prolonging the lifetime of wireless sensor networks by cross-layer interaction. IEEE Wireless Communications 11(6), 78–86 (2004)
11. Madan, R., Cui, S., Lall, S., Goldsmith, A.: Cross-layer design for lifetime maximization in interference-limited wireless sensor networks. In: Proceedings IEEE INFOCOM 2005. 24th Annual Joint Conference of the IEEE Computer and Communications Societies, vol. 3 (2005)

12. Mainwaring, A., Polastre, J., Szewczyk, R., Culler, D., Anderson, J.: Wireless sensor networks for habitat monitoring
13. Hartung, C., Han, R., Seielstad, C., Holbrook, S.: FireWxNet: A multi-tiered portable wireless system for monitoring weather conditions in wildland fire environments. In: Proc. International conference on Mobile systems, applications and services, pp. 28–41 (2006)
14. Werner-Allen, G., Johnson, J., Ruiz, M., Lees, J., Welsh, M.: Monitoring volcanic eruptions with a wireless sensor network. In: Proceeedings of the Second European Workshop on Wireless Sensor Networks, pp. 108–120 (2005)
15. Martinez, K., Ong, R., Hart, J., Stefanov, J.: GLACSWEB: a sensor web for glaciers. Technical University Berlin Telecommunication Networks Group, p. 46
16. ARGO — Global Ocean Sensor Network
17. Tolle, G., Polastre, J., Szewczyk, R., Culler, D., Turner, N., Tu, K., Burgess, S., Dawson, T., Buonadonna, P., Gay, D., et al.: A macroscope in the redwoods. In: Proceedings of the 3rd international conference on Embedded networked sensor systems, pp. 51–63. ACM, New York (2005)
18. Juang, P., Oki, H., Wang, Y., Martonosi, M., Peh, L., Rubenstein, D.: Energy-efficient computing for wildlife tracking: Design tradeoffs and early experiences with zebranet. Computer Architecture News 30(5), 96–107 (2002)
19. Hu, W., Bulusu, N., Chou, C., Jha, S., Taylor, A.: Design and evaluation of a hybrid sensor network for cane toad monitoring (2009)
20. He, T., Krishnamurthy, S., Luo, L., Yan, T., Gu, L., Stoleru, R., Zhou, G., Cao, Q., Vicaire, P., Stankovic, J., et al.: VigilNet: An integrated sensor network system for energy-efficient surveillance. ACM Transactions on Sensor Networks (TOSN) 2(1), 1–38 (2006)
21. Beckwith, R., Teibel, D., Bowen, P.: Pervasive computing and proactive agriculture. In: Advances in Pervasive Computing: A Collection of Contributions Presented at PERVASIVE 2004. Österreichische Computer Gesellschaft, p. 309 (2004)
22. Union, E.: Official journal of the european union
23. Ottman, G., Hofmann, H., Bhatt, A., Lesieutre, G.: Adaptive piezoelectric energy harvesting circuit for wireless remote power supply. IEEE Transactions on Power Electronics 17(5), 669–676 (2002)
24. Joseph, A.: Energy harvesting projects. IEEE Pervasive Computing 4(1), 69–71 (2005)
25. Paradiso, J., Starner, T.: Energy scavenging for mobile and wireless electronics. IEEE Pervasive Computing 4(1), 18–27 (2005)
26. Roundy, S., Leland, E., Baker, J., Carleton, E., Reilly, E., Lai, E., Otis, B., Rabaey, J., Wright, P., Sundararajan, V.: Improving power output for vibration-based energy scavengers. IEEE Pervasive computing 4(1), 28–36 (2005)
27. Williams, C., Yates, R.: Analysis of a micro-electric generator for microsystems. Sensors & Actuators: A. Physical 52(1-3), 8–11 (1996)
28. Raghunathan, V., Kansal, A., Hsu, J., Friedman, J., Srivastava, M.: Design considerations for solar energy harvesting wireless embedded systems. In: IPSN 2005. Fourth International Symposium on Information Processing in Sensor Networks, 2005, pp. 457–462 (2005)
29. Alippi, C., Galperti, C.: An Adaptive System for Optimal Solar Energy Harvesting in Wireless Sensor Network Nodes. IEEE Transactions on Circuits and Systems I: Regular Papers 55(6), 1742–1750 (2008)

30. Duracell, Ni-mh technical bulletin collection, http://www.duracell.com/oem/rechargeable/Nickel/nickel_metal_tech.asp
31. Alippi, C., Galperti, C.: Energy storage mechanisms in low power embedded systems: twin batteries and supercapacitors. In: Proceedings of Wireless Vitae 2009, Aalborg, Denmark, May 17-20, pp. 17–20 (2009)
32. Anastasi, G., Conti, M., Di Francesco, M., Passarella, A.: How to prolong the lifetime of wireless sensor networks. In: Denko, M., Yang, L. (eds.) Mobile Ad Hoc and Pervasive Communications (to appear)
33. Tang, C., Raghavendra, C.: Compression techniques for wireless sensor networks. Wireless sensor networks, 207–231 (2004)
34. Sadler, C., Martonosi, M.: Data compression algorithms for energy-constrained devices in delay tolerant networks. In: Proceedings of the 4th international conference on Embedded networked sensor systems, pp. 265–278. ACM, New York (2006)
35. Madden, S., Franklin, M., Hellerstein, J., Hong, W.: Tag: a tiny aggregation service for ad-hoc sensor networks
36. Boulis, A., Ganeriwal, S., Srivastava, M.: Aggregation in sensor networks: an energy–accuracy trade-off. Ad hoc networks 1(2-3), 317–331 (2003)
37. Goel, S., Imielinski, T.: Prediction-based monitoring in sensor networks: taking lessons from MPEG. ACM SIGCOMM Computer Communication Review 31(5), 82–98 (2001)
38. Cerpa, A., Estrin, D.: Ascent: Adaptive self-configuring sensor networks topologies. IEEE Transactions on Mobile Computing 3(3), 272–285 (2004)
39. Schurgers, C., Tsiatsis, V., Ganeriwal, S., Srivastava, M.: Optimizing sensor networks in the energy-latency-density design space. IEEE transactions on mobile computing (2002)
40. Ganesan, D., Cerpa, A., Ye, W., Yu, Y., Zhao, J., Estrin, D.: Networking issues in wireless sensor networks. Journal of Parallel and Distributed Computing 64(7), 799–814 (2004)
41. Raghunathan, V., Ganeriwal, S., Srivastava, M.: Emerging techniques for long lived wireless sensor networks. IEEE Communications Magazine 44(4), 108–114 (2006)
42. Schott, B., Bajura, M., Czarnaski, J., Flidr, J., Tho, T., Wang, L.: A modular power-aware microsensor with > 1000X dynamic power range. In: Fourth International Symposium on Information Processing in Sensor Networks. IPSN 2005, pp. 469–474 (2005)
43. Tseng, Y., Wang, Y., Cheng, K., Hsieh, Y.: iMouse: an integrated mobile surveillance and wireless sensor system. Computer 40(6), 60–66 (2007)
44. Deshpande, A., Guestrin, C., Madden, S., Hellerstein, J., Hong, W.: Model-driven data acquisition in sensor networks. In: Proceedings of the Thirtieth international conference on Very large data bases. VLDB Endowment, vol. 30, pp. 588–599 (2004)
45. Jain, A., Chang, E.: Adaptive sampling for sensor networks. In: ACM International Conference Proceeding Series, vol. 72, pp. 10–16. ACM, New York (2004)
46. Willett, R., Martin, A., Nowak, R.: Backcasting: adaptive sampling for sensor networks. In: Proceedings of the 3rd international symposium on Information processing in sensor networks, pp. 124–133. ACM, New York (2004)

47. Zhou, J., De Roure, D., Vivekanandan, S.: Adaptive Sampling and Routing in a Floodplain Monitoring Sensor Network. In: Proc. IEEE WiMob, pp. 19–21 (2006)
48. Alippi, C., Anastasi, G., Galperti, C., Mancini, F., Roveri, M.: Adaptive sampling for energy conservation in wireless sensor networks for snow monitoring applications. In: Proc. IEEE International Workshop on Mobile Ad Hoc and Sensor Systems for Global and Homeland Security (MASS-GHS 2007), Pisa, Italy (2007)
49. Raghunathan, V., Schurgers, C., Park, S., Srivastava, M.: Energy-aware wireless microsensor networks. IEEE Signal Processing Magazine 19(2), 40–50 (2002)
50. Vuran, M., Akan, Ö., Akyildiz, I.: Spatio-temporal correlation: theory and applications for wireless sensor networks. Computer Networks 45(3), 245–259 (2004)
51. Pattem, S., Krishnamachari, B., Govindan, R.: The impact of spatial correlation on routing with compression in wireless sensor networks (2008)
52. Vuran, M., Akyildiz, I.: Spatial correlation-based collaborative medium access control in wireless sensor networks. IEEE/ACM Transactions on Networking (TON) 14(2), 316–329 (2006)
53. Akyildiz, I., Melodia, T., Chowdhury, K.: A survey on wireless multimedia sensor networks. Computer Networks 51(4), 921–960 (2007)
54. Alippi, C., Baroni, G., Bersani, A., Roveri, M.: Unsupervised feature selection algorithms for Wireless Sensor Network. In: Proc. IEEE-CIMSA 2009, Hong Kong, China, May 11-13 (2009)
55. Madden, S.R., Franklin, M.J., Hellerstein, J.M., Hong, W.: Tinydb: an acquisitional query processing system for sensor networks. ACM Trans. Database Syst. 30(1), 122–173 (2005)
56. Aberer, K., Hauswirth, M., Salehi, A.: Infrastructure for data processing in large-scale interconnected sensor networks. In: Mobile Data Management (MDM 2007), Mannheim, Germany (2007)
57. Chu, D., Tavakoli, A., Popa, L., Hellerstein, J.: Entirely declarative sensor network systems. In: Proc. VLDB 2006, pp. 1203–1206 (2006)
58. Siemens, Sword - internal communication (2008)
59. Schreiber, F., Camplani, R., Fortunato, M., Marelli, M., Pacifici, F.: PERLA: A Data Language for Pervasive Systems. In: PerCom 2008. Sixth Annual IEEE International Conference on Pervasive Computing and Communications, pp. 282–287 (2008)
60. Beyer, S., Taylor, R., Mayes, K.: Operating system support for dynamic code loading in sensor networks. In: Fourth Annual IEEE International Conference on Pervasive Computing and Communications Workshops. PerCom Workshops 2006, p. 5 (2006)
61. Egea-Lopez, E., Vales-Alonso, J., Martinez-Sala, A., Pavon-Marino, P., Garcia-Haro, J.: Simulation scalability issues in wireless sensor networks. IEEE Communications Magazine 44(7), 64 (2006)
62. The Network Simulator - ns-2
63. Naoumov, V., Gross, T.: Simulation of large ad hoc networks. In: Proceedings of the 6th ACM international workshop on Modeling analysis and simulation of wireless and mobile systems, pp. 50–57. ACM, New York (2003)

64. Sobeih, A., Chen, W., Hou, J., Kung, L., Li, N., Lim, H., Tyan, H., Zhang, H.: J-sim: A simulation environment for wireless sensor networks. In: Proceedings of the 38th annual Symposium on Simulation, pp. 175–187. IEEE Computer Society, Washington (2005)
65. Varga, A., et al.: The OMNeT++ discrete event simulation system. In: Proceedings of the European Simulation Multiconference (ESMí 2001), pp. 319–324 (2001)
66. Levis, P., Lee, N., Welsh, M., Culler, D.: TOSSIM: Accurate and scalable simulation of entire TinyOS applications. In: Proceedings of the 1st international conference on Embedded networked sensor systems, pp. 126–137. ACM, New York (2003)
67. Polley, J., Blazakis, D., McGee, J., Rusk, D., Baras, J.: Atemu: A fine-grained sensor network simulator. In: 2004 First Annual IEEE Communications Society Conference on Sensor and Ad Hoc Communications and Networks. IEEE SECON 2004, pp. 145–152 (2004)
68. Girod, L., Elson, J., Cerpa, A., Stathopoulos, T., Ramanathan, N., Estrin, D.: Emstar: a software environment for developing and deploying wireless sensor networks
69. Alippi, C., Camplani, R., Galperti, C., Roveri, M., Sportiello, L.: Towards a credible WSNs deployment: a monitoring framework based on an adaptive communication protocol and energy-harvesting availability. In: IEEE Instrumentation and Measurement Technology Conference Proceedings, 2008. IMTC 2008, pp. 66–71 (2008)
70. Al-Karaki, J., Kamal, A.: Routing techniques in wireless sensor networks: a survey. IEEE Wireless Communications 11(6), 6–28 (2004)
71. Akkaya, K., Younis, M.: A survey on routing protocols for wireless sensor networks. Ad Hoc Networks 3(3), 325–349 (2005)
72. P2ict lab monitoring the marine environment, http://www.prometeo.polimi.it/ict/icteng/ict_australia_eng.html
73. XBow, http://www.xbow.com/
74. MaxStream, http://www.maxstream.com/
75. International Society of Rock Mechanics, Suggested methods for monitoring rock movements using inclinometers and tiltmeters. Rock Mechanics 10, 81–106
76. Lynch, P.: An overview of wireless structural health monitoring for civil structures, ser. A. In: Philosophical Transactions of the Royal Society of London. A Mathematical and Physical Sciences. The Royal Society, London (2005)
77. Green, L.G., Maürer, H., Spillmann, T., Heincke, B., Willenberg, H.: High-resolution geophysical techniques for improving hazard assessment of unstable rock slopes. Swiss Federal Institute of Technology, Zurich
78. Alippi, C., Camplani, R., Galperti, G.: Lossless compression techniques in wireless sensor networks: Monitoring microacoustic emissions. In: IEEE International Workshop on RObotic and Sensors Environments, ROSE 2007, Ontario, Canada, October 12-13, pp. 1–5 (2007)

Loading Analysis of a Remotely Interrogatable Passive Microvalve

Ajay C. Tikka[1], Said F. Al-Sarawi[1], and Derek Abbott[2]

[1] Centre for High Performance Integrated Technologies and Systems (CHiPTec), The University of Adelaide, SA 5005, Australia
ajay.tikka@adelaide.edu.au
[2] Centre for Biomedical Engineering, The University of Adelaide, SA 5005, Australia

We present the dynamic loading analysis of a normally closed, remotely actuated, secure coded, electrostatically driven, active microvalve using passive components. The design employs a synergetic approach to incorporates the advantages of both electroacoustic correlation and electrostatic actuation into the microvalve structure. This is carried out by utilising the complex signal processing capabilities of two identical, 5×2-bit Barker sequence encoded, acoustic wave correlators. An electrostatically driven microchannel, comprising of two conducting diaphragms as the top and bottom walls, is placed in between the compressor IDT's of the two correlators. Secure interrogability of the microvalve is demonstrated by the 3-D finite element modelling of the complete structure and the quantitative deduction of the harmonic code dependent microchannel actuation. Furthermore, the dynamic transient analysis is employed to investigation the nonlinear time response of the microvalve and other performance criteria of the structure such as microchannel opening dynamics and the microvalve loading time.

1 Introduction

With the ever increasing demand for high performance microfluidic devices and continuous quest for miniaturisation, the development of novel and innovative microvalves has witnessed a rapid growth in the recent times. Most of the research in this area is focused on improving the performance of the microvalve by targeting features such as flow rate, leakage flow, power consumption, response time and disposability [1, 2]. However, the constant expansion of the application domain of these devices into fields as diverse as life sciences and chemistry applications is placing new demands on the existing models. One among these new design requirements is the remote interrogability of the microfluidic devices. The capability to wirelessly control fluid flow can emerge as an attractive technology enabling various biomedical applications. Furthermore, most of the existing microfluidic devices in biomedical implants are either battery powered or have external pump modules that are connected to the patient by means of a passive port [3, 4]. The use of a battery would place additional constraints on the device size and would require an electronic circuitry module. Hence, the development of an active microvalve with fully passive components, enabling wirelessly control of fluid flows, is a key task in the realization of compact high performance biomedical implants. Moreover, such a device would provide a freely programmable, time-modulated control profile to ensure patient safety and comfort in long-term treatment.

Microfluidic devices using acoustic streaming phenomenon have been investigated in the past [5, 6]. These devices use surface acoustic waves or flexural plate waves to cause fluid motion, through hydrodynamic coupling, on the planar surface of the piezoelectric material. Acoustic streaming based microfluidic devices have several advantages such as low fluidic impedance of the channel, simple structure and low sensitivity to the electrical and chemical properties of the fluid. However, secure actuation of the wireless devices using acoustic streaming is limited only to frequency addressability rather than discrete code addressability, thus rendering them inefficient for remote interrogation applications. This shortcoming is tackled, in this paper, by using two identical 5×2-bit Barker sequence encoded surface acoustic wave (SAW) correlators placed one on top of the other with an air gap and suspending an edge clamped microchannel with two conducting diaphragms between the compressor interdigital transducer's (IDT) of the correlators. The correlator's response is employed to impart ultrasonic energy on the conducting diaphragms using electrostatic actuation. Thus combining the high frequency operation of the ultrasonic, secure code embedded, acoustic correlator with low power, fast response time, and reliability of electrostatic actuation. Successful wireless operation of this device hinges on the precise occurrence of fluid flow when interrogated by a correlating BPSK signal. The microvalve is normally closed and is opened by using the deformation of the diaphragms caused by the electrostatic effect. The risk of high leakage and valve clogging is addressed by employing both diffuser elements and check valves in the structure.

Small device dimensions make it typically complicated or even impossible to perform measurements of the physical conditions inside microdevices. To facilitate optimal design of MEMS devices based on such microactuators, efficient simulation schemes for analysing the channel deformation are required. The physical behaviour of micro components is best described by partial differential equations, which are typically solved by finite element methods. Due to the complexity involved in combining the electroacoustic and electrostatic mechanisms, a systematic design and optimization of the active microvalve is best possible with finite element method (FEM) modelling. Furthermore, coupled field simulations are vital for capturing both electroacoustic and electrostatic-structural interactions in a single finite element run and for taking into account non-linear effects that are inherent in MEMS devices. Hence, in this work, we model the whole structure by employing a direct FEM to verify the functionality of the concept. The FEM simulation architecture for the structural components of the microvalve consists of two major subsystems: a SAW Correlator electroacoustic analysis, including the voltage response and the influence of the excited acoustic modes, as was discussed in the previous publication [7], and a microchannel electrostatic analysis that captures the channel deformation and the code dependent behaviour of the structure.

In this paper, the design and simulation of a 3-dimensional, remotely interrogatable, active microvalve is presented using two 5×2-bit Barker sequence encoded SAW correlators. In Section 2, the use of SAW devices for microfluidic applications including acoustic streaming is outlined. In Section 3, we discuss briefly the principle of operation, the transmitter receiver architecture, electrostatic actuation and the finite element formulation. Then in Section 4, we present the microchannel electrostatic model, where the dimensions of the structure and materials are discussed. In section 4, the results of the comprehensive harmonic and transient deflection analysis of the microvalve for variable acoustic modes and different input interrogating code sequences are presented.

2 SAW Devices for Microfluidic Applications

In recent times extensive research and development efforts are globally directed towards fluid diagnostics and flow manipulation by adopting advances made in other expansive fields such as MEMS, micromachining and material sciences. The novel miniature microfluidic pumps, valves, sensors and controllers have reached a level of maturity to make deep inroads in to both civilian and military markets. As per market studies, the microfluidic component market is projected to reach $4.2 billion by 2014 (Micro Fluid Management Technology - Yole Development). It is also reported that the trend towards function integration would drive the demand for next generation of microfluidic applications such as portable medical devices, micro fuel cells, and micro chemical reactors. The use of SAW technology for such applications has the potential to open up many new possibilities. The main advantage is that the SAW devices, being inherently passive and robust, are not severely impacted by the hostile conditions. Furthermore, with the ability to wirelessly power and control these devices they can be placed in remote locations.

The SAW devices are commonly used in consumer electronics as high frequency filters and signal processing components. Over the past two decades the use of SAW devices for sensing, especially for microfluidic applications, has gained momentum. To be quantified as a sensor a sensing layer, which is sensitive to the input stimuli, is deposited in between or above the IDT's. The variation in the properties of the sensing layer when contacted by the fluid determines the dynamics of the interaction between the charges at or close to the surface of the sensing layer and the underlying SAWs. This is measured in terms of deviation of the wave velocity, phase angle or the attenuation of the output response of the SAW sensor. The acoustic wave sensors have the additional capability to sense both gas and liquid media. The development of a wide array of acoustic sensors have been reported in the literature targeting microfluidc sensing quantities such as gases at high temperature [8], viscosity [9], humidity [10], antibody-antigen reactions [11], DNA molecules [12], and protein samples [13].

2.1 Acoustic Streaming

Recently SAWs are increasingly being employed to act in a completely different way than for sensing in micro-fluidic applications, i.e., for the manipulation of fluid flow. This is carried out by a phenomenon known as acoustic streaming where the interaction of the acoustic wave with the fluid on the surface, as shown in the Fig. 1, induces an internal streaming. This internal streaming causes a droplet motion in the range of microliters down to picoliters. As most of the energy propagating in a SAW is of mechanical nature the acoustic radiation pressure exerted on the fluid drives it in the direction of the propagation of the acoustic wave. Fluid tracks are created on the surface of the substrate, to confine the liquid flow to a predetermined path, by the chemical variation of the surface tension of the tracks. The acoustic streaming technique reduces the reliance on other passive flow control mechanisms, such as capillary force, needed to overcome the high impedance of the channel.

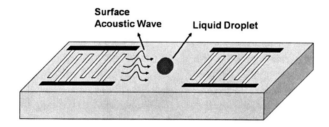

Fig. 1. Acoustic Streaming of liquid using a SAW device

The ridges and wells of the acoustic wave create a fluid pressure difference ($2\delta p$) which leads to a difference in the liquid density ($2\delta \rho$). Thus, the excited longitudinal acoustic wave, into the liquid medium, comprises of oscillating pressure and density components with an equilibrium values of p_0 and ρ_0. If v_S and v_L are the acoustic velocities in the piezoelectric substrate and liquid medium respectively then the diffraction angle of the launched wave is given by [14, 6]

$$\Theta_D = \arcsin(\frac{v_S}{v_L}). \qquad (1)$$

Furthermore, the radiation pressure of the launched wave that causes the internal streaming of the liquid is given by

$$p_a = \rho_0 v_S^2 (\frac{\delta \rho}{\rho_0})^2. \qquad (2)$$

Many microfluidic devices based on acoustic streaming such as micropumps [15, 16], microliquid heating systems [17], micro-particle removal devices [18], and ultrasonic motors [19] have been reported in the literature.

These devices thrive on the numerous advantages offered by the SAW technology. Some of them are simple device structure, ease of manufacturing, high reliability due to the lack of moving components, and improved channel resistance because of the presence of propelling force throughout the fluidic network. Moreover, very large forces are generated over a very small area due to the scalable nature of the acoustic streaming phenomenon. This enables small scale applications where the contribution of acoustic force to induced streaming is significant. However, the acoustic steaming based microfluidic devices are plagued by their share of limitations as well. These devices suffer from low efficiency as the induced streaming is a second order, nonlinear phenomenon. A plausible way of addressing this shortcoming is by employing large amplitude acoustic waves, which would make streaming inappropriate for applications where power usage is critical such as wireless, battery-less applications. Moreover, the amount of heat generated at the solid liquid interface for such high acoustic amplitudes is detrimental for thermally sensitive liquids. Due to these limitations, to the best of authors knowledge, no acoustic streaming based wireless, fully passive microfluidic device has been reported so far. Even if developed, by employing recent innovations such as focusing of acoustic power with horns, lenses and focused transducers [20, 21], the secure actuation of such a device is limited only to frequency addressability rather than discrete code addressability, which would render it inefficient for remote interrogation applications. This constraint is resolved in this research by employing a a synergetic approach which would combine electroacoustic correlation and electrostatic mechanisms, as will be discussed in detail in the next section.

3 Wireless Microvalve Design

The microvalves can be classified in two broad categories namely active and passive. While the the fluid flow in passive microvalves is controlled by the fluid pressure the active microvalves are triggered by an external signal irrespective of the fluid dynamics. Active microvalves have emerged as the predominant form of microfluidic components currently being employed for a wide array of applications ranging from on/off switching to fluid flow control and isolation. Innovations in this area are normally aimed at developing novel devices with optimising characteristics such as valve closing dynamics, response time, power consumption, leakage, flow resistance and reusability. However, not much research effort is directed towards remotely controlled flow manipulation, which has the potential to pave way for the next generation of microfluidic devices. The possible applications of a remote controlled, fully passive microvalve are boundless, especially for drug delivery applications in the biomedical field. The ability to wirelessly administer accurate doses of drug, for an extended period of time, at an inaccessible target location, through an implanted microvalve can revolutionise the way diseases such as cancer and diabetes are treated.

The design of a normally closed, remotely actuated, secure coded, batteryless, active microvalve using fully passive components is presented in this chapter. The crucial choice of actuation mechanism for microvalves is normally guided by the design specifications. Even though acoustic streaming appears to be a suitable actuation principle, that comes closest to satisfying all the above mentioned design specifications, it falls short in addressing the power and secure coding requirements. These drawbacks are alleviated in this research by combining the complex signal processing capabilities of acoustic wave correlator with the electrostatic actuation of the microchannel. This synergetic approach incorporates the advantages of both the mechanisms into the microvalve design.

3.1 Principle of Operation

The design of the current novel, passive, wireless microvalve is motivated by the aim of high-resolution volumetric dosing and the demand for a minimum device size and power consumption. The core components of the microvalve are two identical SAW Correlators, two diaphragms and an antenna array. The diaphragms are suspended with an air gap between the compressor IDT's of the correlators forming a microfluidic channel. The resulting correlated output at the compressor IDT of the correlator generates a electrostatic field within the air gap which in turn deflects the diaphragms, thereby regulating fluid flow through the microfluidic channel.

The wireless telemetry system for the SAW based microvalve is depicted in Fig. 2. By means of the antennas, the microvalve captures part of the electromagnetic energy to provide power for its own operation. One of the

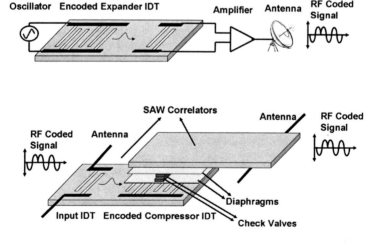

Fig. 2. Wireless microvalve transmitter receiver configuration

advantages of this approach is the ability to establish bidirectional communication between the microvalve and the interrogator. Moreover, this concept does not allow simultaneous triggering of several microvalves present in the same electromagnetic field as each device is encoded with a different code. System level protocols, such as those in use in contactless smart-card readers, allows a single controller to communicate with several individually identified microvalve.

A coded SAW based communication system for the microvalve, consisting of an expander IDT in the transmitter and a compressor IDT in the receiver, is briefly outlined in this paper. A narrow pulse or a sinusoidal waveform is fed to the expander IDT to generate a coded acoustic signal depending on the geometry of the expander IDT. These acoustic waves propagate through the substrate to the transmitting IDT, which transforms these coded acoustic waves to electrical coded RF signal. The output from the transmitting IDT is fed to an amplifier, to strengthen the signal, and then to a transmitting antenna. The receiver consists of two identical correlators, the operation of which is explained in the next section, with their input IDT's connected to the receiving antennas to intercept the transmitted coded RF signal. The expander in the transmitter is an exact replica of the compressor/coded IDT of the correlators. The coding of the expander and compressor determines the autocorrelation function performed by the correlators. The resultant response of the acoustic devices is then electrostatically coupled to the conducting diaphragms, suspended between the compressor IDT's with an air gap. Thus the actuation method not only depends on the excitation frequency but also on the code of the transmitted BPSK signal. The microchannel comprises of two diaphragms with check valves and an inlet and outlet diffuser element on the either side of the channel. Apart from the fluid flow resistance of the inlet and outlet diffuser elements, the leakage is further controlled by the check valves. When an electrostatic force is applied, the diaphragms are pulled towards the compressor IDT of the correlators causing the check valves to separate. Thus, the simultaneous actuation of the diaphragms at ultrasonic frequencies by the correlating input signal allows the microvalve to induct and expel fluid through the check valves in the channel.

The valve efficiency is a critical parameter for microvalves used in on/off switching applications. The valve efficiency of an active microvalve with diffuser elements is poor in the reverse direction due to high leakage. The desired leak-tight operation makes it necessary to incorporate micro check valves on the diaphragms in the fluid channel. The lateral dimensions of the check valves are designed to make them stiff enough to resist bowing under pressurization caused by the fluid in the idle state and yet compliant enough to allow for the diaphragm motion during actuation. The sketches Fig. 3 depict the functionality of the microvalve in the OFF/normally closed state and ON state. In the OFF state the check valves push tightly against each other as shown in Fig. 3(a). Assuming a perfect check valve, the fluid flow through the channel is shut-off in the OFF state or even when there is a mismatch in the

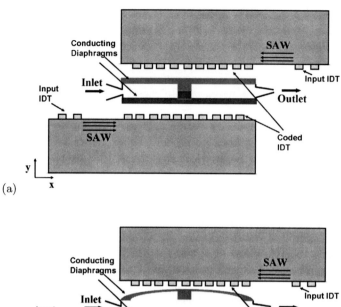

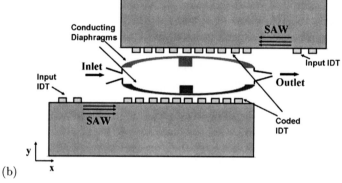

Fig. 3. Microvalve in (a) OFF/Normally closed state. (b) ON state.

operating frequency and code of the interrogating BPSK signal. In the ON state, when interrogated by a correlating signal the double membranes inflate due to electrostatic actuation and inhale the fluid into the chamber. Since this is happening at ultrasonic frequency the expansion and compression of the channel volume allow the compressed fluid to escape through the outlet. The microchannel dimensions and the diaphragm deflection depend on the application envisaged for the device. The piezoelectric material's electromechanical coupling capabilities, the encoded code of the SAW correlator, the diaphragm material and thickness, the air gap, the compliances of the fluid and check valve elements in the channel, and the nature of the interrogating RF signal all contribute to the performance of the microvalve device.

The fabrication of microfluidic devices is time consuming, expensive and complex. The lack of a generalised architecture, especially for a specialised design such as the one proposed in the current research, necessitate a thorough understanding of the multiple coupled physics phenomenon at the process level, before fabrication. Moreover, the measurement of physical quantities in miniature microdevices is complicated and in some instances unfeasible. However, with the use of FEM based multiphysics modelling the interactions between electromagnetic, piezoelectric, electroelastic and fluid fields can be analysed to gain a deeper insight into the operation and functioning of microdevices. The FEM multiphysics provides a rapid and inexpensive design evaluation, geometry independent resolution and the ability to seamlessly incorporate multiple materials and different loading conditions. Furthermore, a comprehensive nonlinear transient loading and dynamic frequency characteristics of the microfluidics can be obtained with the inclusion of parameters such as structural spring force, squeeze film damping, fringing field, and intrinsic residual stresses, to simulate the test conditions.

In the current research the emphasis is on the design and optimisation of the novel microfluidic structure through the deflection analysis, both, to verify the functionality of the concept and to investigate the working range of the structure. Hence, the primary focus is on secure actuation caused by combining the electroacoustic and electrostatic mechanisms. The optimisation of the microchannel design is addressed by the consideration of the diaphragm materials with enhanced mechanical and surface properties, and rational and reproducible selection of air gap while integrating the SAW correlator with the microchannel. The contributions of electroacoustic correlation and electrostatic mechanisms to the microvalve design are analysed in detail in the following subsections.

3.2 Electroacoustic Correlation

The need to optimise and verify individual components of the device in order to achieve the desired performance of the integrated assembled system is clearly evident. The FEM modelling, as was discussed in the previous publication [7], and experimentally validation of SAW correlator was considered first. The use of SAW technology in general and correlation in particular bestows several benefits to the microvalve structure. The contributions of SAW filter technology include small size, passivity, low cost, reliability, robustness, compact packaging, relatively stable performance over temperatures, ease of device fabrication for both low and high volume production. Furthermore, the use of SAW correlation provides remote discrete code addressability, low loss and sharp signal suppression from resonance to rejection band (high Q-factor). The next step would be to impart the measured correlator's response on the microchannel and analyse the electrostatic actuation.

3.3 Electrostatic Actuation

Electrostatic mechanism is one of the preferred methods of actuation for microfluidic devices due to fast response time, low power consumption, ease of integration, compatibility with other fabrication technologies and independency on the ambient temperature [22, 23]. The electrostatic actuators

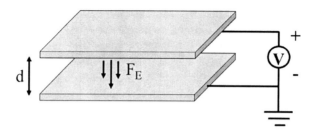

Fig. 4. Parallel plate electrostatic actuator

utilise the induced forces between closely placed conducting plates/electrodes that are energised to different potentials, as shown in the Fig. 4. Using basic electrostatics, this force can be written as

$$F_E = \frac{\epsilon A_o}{2(d - \mathbf{a})} V^2. \tag{3}$$

Where, ϵ is dielectric constant of the plate material, A_o the overlapping area of the plates, V is the applied potential difference, d is the initial separation between the plates and $\mathbf{a}$ is the displacement of the moving plate from its initial equilibrium. This electrostatic force is significant in the micro scale and can deform one of the plates until they are balanced by the restoring elastic forces. Apart from the plate thickness and material type, the quadratic increase in the induced electrostatic force for decreasing distance between the plates determines the range of deflection. With the recent advances in micromachining techniques the two plates can be placed very close each other to generate large field strength for moderate plate potentials. This is crucial for the current design where the wireless powering imposes tighter constraint on the voltage available for microvalve actuation. However, by varying the transmitted power levels the deflection range and therefore the ejected fluid quantity from the microvalve can be controlled.

Finite Element Formulation

The need for the careful design and optimisation of the electrostatically driven microchannel structure has lead to the use of FEM modelling. The finite element design formulation along with the ingrained differential equations are discussed here. The coupled field microchannel model is descretized into finite elements and solved by a three-field formulation. Here, in addition to use of structural elements and electrostatic elements for describing the conducting diaphragms and air gap respectively, a third field is introduced to account for the motion of the electrostatic mesh due to geometric and material nonlinearities.

According to electromechanical energy conservation principle [24], the net flow of energy into the lossless system, due to applied loads, is offset by the internal change of energy stored in the system. Thus, the expression for equilibrium structural solution can be written as [25, 26]:

$$\mathbf{F} = [M]\frac{\partial^2 \mathbf{u}}{\partial t^2} + [C]\frac{\partial \mathbf{u}}{\partial t} + [K]\mathbf{u}, \qquad (4)$$

here, the mechanical force vector $\mathbf{F}$ comprises of two components, the nodal forces applied to the element and $\mathbf{F}^N$ and the thermal force component $\mathbf{F}^{TH}$ given by

$$\mathbf{F}^{TH} = \int_{vol} [G]^T [D] \epsilon^{TH} d(vol), \qquad (5)$$

where $[G]$ is the strain-displacement matrix determined by the element shape. $[M]$ is the mass matrix given by

$$[M] = \rho \int_{vol} [N]^T [N] d(vol), \qquad (6)$$

where $[N]$ is structural shape function of the element and ρ is the density of the material used for the microchannel. $[K]$ is structural stiffness matrix expressed as

$$[K] = \int_{vol} [G]^T [D][G] d(vol), \qquad (7)$$

where $\mathbf{D}$ is the electric flux density vector in the electrostatic domain. If $[N_E]$ is electrical shape function of the element then the dielectric permittivity matrix $[K_d]$ is

$$[K_d] = -\int_{vol} [N_E]^T \epsilon [N_E] d(vol). \qquad (8)$$

The dynamic equilibrium equation for an electrostatically actuated microchannel is influenced by the dissipated or stored mechanical energy of the structural elements and the stored electrostatic energy of the electrostatic elements. A strongly coupled tetrahedral electrostatic element is used in this work to accurately model the electro-mechanical coupling. Based on the developed model geometry and design requirements, the structural equilibrium expression in Eqn. (4) can be modified to represent a electrostatic system:

$$\begin{pmatrix} M & 0 \\ 0 & 0 \end{pmatrix} \begin{pmatrix} \frac{\partial^2 \mathbf{u}}{\partial t^2} \\ \frac{\partial^2 \mathbf{v}}{\partial t^2} \end{pmatrix} + \begin{pmatrix} C & 0 \\ 0 & 0 \end{pmatrix} \begin{pmatrix} \frac{\partial \mathbf{u}}{\partial t} \\ \frac{\partial \mathbf{v}}{\partial t} \end{pmatrix} + \begin{pmatrix} K & 0 \\ 0 & K_d \end{pmatrix} \begin{pmatrix} \mathbf{u} \\ \mathbf{v} \end{pmatrix} = \begin{pmatrix} \mathbf{F} + \mathbf{F}^E \\ \mathbf{L} \end{pmatrix}. \qquad (9)$$

Here, the nodal electrostatic force $\{F^E\}$ that contributed to the microchannel deformation can be expressed as

$$\{F^E\} = \int_{vol} [G]^T \sigma^M d(vol), \qquad (10)$$

where σ^M is the Maxwell stress vector obtained from

$$[\sigma^M] = \frac{1}{2}(\mathbf{E_k}\mathbf{D}^T + \mathbf{D}\mathbf{E_k}^T - \mathbf{D}^T\mathbf{E_k}[I]), \tag{11}$$

$\mathbf{E_k}$ and $\mathbf{D}$ are the electric field intensity and electric flux density vectors, respectively, in the electrostatic domain and $[I]$ is the identity matrix. For the design analysis performed in this research, the complete solution is obtained via the solution of equation Eqn. (9).

As the electrostatic actuation involves the coupling of electrostatic and elastomechanic fields, any deformation in the microchannel would require a remeshing of the structure due to the nonlinear structural stiffness of the clamped diaphragms. This is carried out by using Newton-Raphson method [27] where the non-linear system is iteratively solved for convergence using finite-differences.

$$\begin{pmatrix} \frac{\partial \mathbf{R}_s^i}{\partial \mathbf{u}} & \frac{\partial \mathbf{R}_s^i}{\partial \mathbf{v}} \\ \frac{\partial \mathbf{R}_e^i}{\partial \mathbf{u}} & \frac{\partial \mathbf{R}_e^i}{\partial \mathbf{v}} \end{pmatrix} \begin{pmatrix} \Delta \mathbf{u}^i \\ \Delta \mathbf{v}^i \end{pmatrix} + \begin{pmatrix} \mathbf{R}_s^i \\ \mathbf{R}_e^i \end{pmatrix} = 0. \tag{12}$$

Here $\mathbf{R_s}$ and $\mathbf{R_e}$ are discretized residual vectors of elastostatic and electrostatic field respectively for a nodal charge density of q_n, given by

$$\mathbf{R}_s(\mathbf{u}, \mathbf{v}) = \mathbf{F}(\mathbf{u}) - \mathbf{F}^E(\mathbf{u}, \mathbf{v}), \tag{13}$$

and

$$\mathbf{R}_e(\mathbf{u}, \mathbf{v}) = \epsilon_e(\mathbf{u}, \mathbf{v}) - q_n. \tag{14}$$

The updated values of $\Delta \mathbf{u}^i$ and $\Delta \mathbf{v}^i$ for the ith iteration in Eqn. (12) are calculated from

$$\begin{pmatrix} \mathbf{u}^{i+1} \\ \mathbf{v}^{i+1} \end{pmatrix} = \begin{pmatrix} \mathbf{u}^i \\ \mathbf{v}^i \end{pmatrix} + \begin{pmatrix} \Delta \mathbf{u}^i \\ \Delta \mathbf{v}^i \end{pmatrix}. \tag{15}$$

For the solution to converge Eqn. (12) and Eqn. (15) are solves repeatedly until both residual vectors approach zero. The FEM modelling of the microchannel is discussed in the next section.

4 Microchannel Electrostatic Actuation

The fast response times and the absence of friction in the electrostatic actuators make them an appropriate choice for high speed ultrasonic applications. However, the usage of electrostatic actuation to achieve moderate displacements is challenging in a wireless environment. This is mainly due to the tradeoff between the applied voltage and the gap between the electrodes. If the air gap between the diaphragm and the compressor IDT is increased then the required amplitude of the interrogating signal to actuate the microvalve becomes extremely high. On the other hand, if the air gap is small, the diaphragm might run the risk of shorting with the compressor IDT. Therefore, the usage of two identical SAW correlators with two diaphragms is a way of addressing this problem and hence optimising the microchannel deflection. As both the diaphragms are grounded the lack of electrical connections near or inside the channel provides electrical isolation of the fluid and simplifies channel sealing.

In the current microchannel model, as shown in the Fig. 5, the actuating field is between a 2 μm thick polyimide diaphragms and the output IDT of the correlators, separated by a 1.5 μm air gap. A biocompatible Polyimide is chosen as the diaphragm material due to its high conductivity and low Young's modulus constant. The length of the edge clamped diaphragms (560 μm) is determined by the length of the output IDT of the correlators. Short diaphragm lengths consume less die area and increase the actuator resonant frequency. Longer diaphragms, however, reduce operating voltage requirements [28]. While design rules limit how close a diaphragm can be to a SAW correlator, manufacturability issues limit the thickness of the diaphragm. The width of the microchannel is taken as 0.5 mm conforming with the overlapping area between the diaphragms and the compressor IDTs of the correlators. The flexible electrode or conducting diaphragm is electrically connected to ground, while the voltage at the fixed electrode or output IDT of the SAW correlator is determined by the input interrogating BPSK signal. In this work, the measured output voltage response of a 5×2 Barker sequence SAW correlator, is employed to drive the microchannel. The electrostatic force is generated due to the electric field between the interdigitated fingers and the diaphragms. Thus, this attractive force results in the center deflection of the diaphragms towards the compressor IDTs of the correlators.

The demands for accurate modelling of the structure, replicating the physical conditions, and the strong dependence of microchannel deflection on the width of the diaphragm prompted the use of 3-D modelling. A meshed 3-D microchannel structure along with the compressor IDTs of the correlators is depicted in the Fig. 6. Tetrahedral elements with four degrees of freedom, three for displacement and one for potential, are used to model the structure.

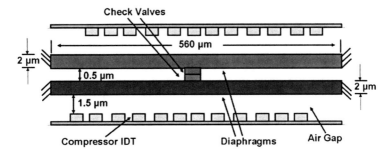

Fig. 5. Microchannel electrostatic model

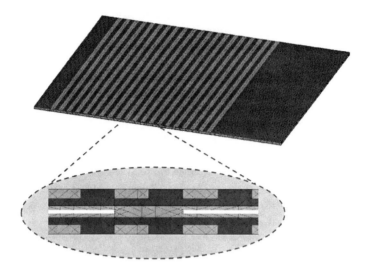

Fig. 6. The meshed microchannel structure with the check valves in the inset

5 Results

The main objective in the evaluation of the microvalve is to measure the center deflection of the microchannel when the correlator is interrogated with a matched and mismatched BPSK signal. This key parameter determines the working range of the microvalve. The interrogating code and acoustic mode dependent behaviour of the microvalve is studied using harmonic analysis. Furthermore, the displacement and the potential contours are employed to get a deeper insight into the microvalve actuation. In addition to analysing the nonlinear time response of the microvalve, the dynamic transient analysis assists in the investigation of other performance criteria of the structure such as microchannel opening dynamics and the microvalve loading time. This is performed by looping the electrostatic solution and the structural deformation till the convergence is attained, as outlined by the iterative Eqn. (12) and Eqn. (15).

A fabricated 5×2 Barker sequence correlator is driven by a ±10V BPSK signal and the measured voltage response at the compressor IDT is utilised for the microvalve design. Fig. 7(a) shows the center deflection of both the diaphragms in the y direction, when the code matches, for different acoustic modes determined by the excitation frequency. Substantial microchannel displacement can be observed in the vicinity of the excited modal frequencies of the correlator. Hence, the electrostatic analyses of the microchannel for a widely varying range of frequencies acts as an effective method for displacement optimisation. This allows us to pick the operating mode of interest for displacement optimisation. From Fig. 7 (b), a maximum separation of the check valves of 0.59 μm

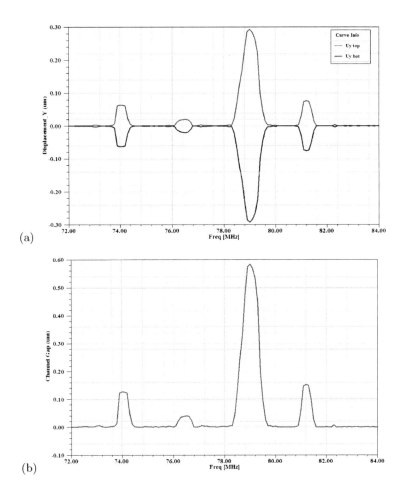

Fig. 7. The center deflection of the diaphragms when the code matches (a) In the y direction. (b) The channel gap between the check valves.

can be observed at the SAW mode at 79 MHz. The y-displacement contour and the electric potential contour of the check valves at the same modal frequency is given in Fig. 8 to analyse the microchannel opening. Here, the separation between the check valves can be observed. Even though this displacement is small compared to thermal and piezoelectric bimorph actuators, the high frequency operation of the microvalve results in high particle velocities [29]. The diaphragm deflection can be further increased by employing corrugated diaphragm structures or by considering correlators encoded with a longer code, which in turn increasing the diaphragm length. This could not be implemented in the current work due to the huge constraint imposed by such a microvalve modelling technique on the available computational resources.

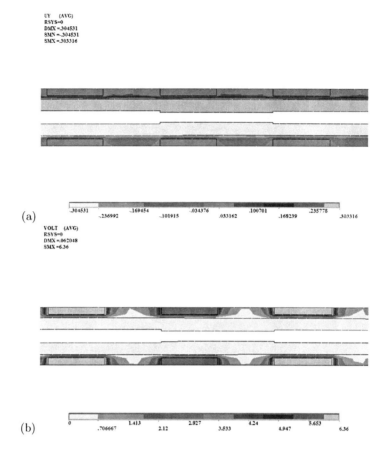

Fig. 8. Contour plots of the check valves at SAW modal frequency. (a) Y-displacement contour. (b) Electric potential contour.

In addition to the harmonic analysis, a dynamic transient analysis of the microchannel is carried out to develop a complete nonlinear finite element model. With this analysis, the transient microchannel deflection and the microvalve loading time are derived by interrogating the microvalve with a continues sequence of matched BPSK signal at SAW modal frequency. Fig. 9 depicts the transient center deflection of both the diaphragms in the y-direction for a period of 3.5 microseconds. The deflection is very low initially due to the time taken by the surface acoustic wave to propagate from the input IDT to the compressor IDT of the correlator. Moreover, the mismatched filling of the correlator caused by the non-synchronisation between the coded acoustic wave and the encoded electrodes of the compressor IDT contributes to this delay. From the instance when the synchronous filling of the correlator occurs the microvalve stabilises, as can be observed from the Fig. 9 is after 650

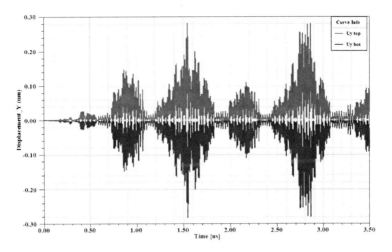

Fig. 9. Acoustic Streaming of liquid using a SAW device

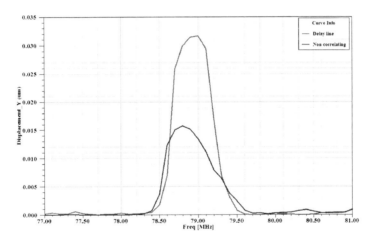

Fig. 10. The channel gap between the check valves when the code mismatches for a delay line input and non-correlating input

ns. After this time the electrodes of the compressor IDT pick up the coded acoustic waves synchronously and hence a consistent deflection peaks can be observed from then on.

So far the results are confined to the instance when there is a code match between the input BPSK signal and the compressor IDT of the correltor. The functionality of the microvalve can only be verified by observing the microchannel deflection to a mismatched input code. This is carried out by varying the electrode coupling of the expander IDT. The response of the microchannel to two different codes, a delay line input and other non-correlating input, at the SAW mode of interest is provided in Fig. 10. All the other specifications of the model are kept the same except for the mismatched input code.

By comparing these responses with the correlating microchannel deflection response of the Fig. 7 it can be established that in addition to the excitation frequency, the microvalve's response is determined by the code in the RF signal. Even in the case of a delay line input, when the two codes are very similar, the device response is much less than its response to the matched code, as can be seen from Fig. 10.

6 Conclusion

The paper presented the modelling and simulation of a novel wireless active microvalve by taking into account both the electroacoustic and electrostatic mechanisms in a single finite element run. The requirements placed by the current and emerging biomedical applications on these devices such as small size, passivity, and remote interrogability are taken into consideration. The results discussed in this work include: (i) comprehensive FEM formulation of 3-Dimensional, 5×2-bit Barker sequence SAW correlator driven microchannel (ii) detailed study of the harmonic and dynamic transient microchannel deflection cause by electrostatically coupling the output IDT's of the SAW correlators to the microchannel and analysis of difference in microchannel deflection when there is a code mismatch. It is concluded that the model appropriately represented the interrogating signal code dependent, and the excited acoustic mode dependent operation of the microvalve and hence enabled the analysis of the microchannel deflection. In consideration of such actuation analysis by the developed FEM model, a microfluidic device with more optimal performance is expected to be designed for various applications.

References

1. Oh, K.W., Ahn, C.H.: A review of microvalves. Journal of Micromechanics and Microengineering 16, 13–39 (2006)
2. Nguyen, N., Huang, X., Chuan, T.K.: MEMS-Micropumps: A Review. Journal of Fluids Engineering 124(4), 384–392 (2002)
3. Cao, L., Mantell, S., Polla, D.: Implantable medical drug delivery systems using microelectromechanical systems technology. In: Proc. of 1st International Conference on Microtechnologies in Medicine and Biology, October 2000, pp. 487–490 (2000)
4. Geipel, A., Doll, A., Goldschmidtböing, F., Müller, B., Jantscheff, P., Esser, N., Massing, U., Woias, P.: Design of an implantable active microport system for patient specific drug release. In: BioMed 2006: Proceedings of the 24th IASTED International Conference on Biomedical Engineering, February 2006, pp. 161–166 (2006)

5. Demirci, U.: Acoustic picoliter droplets for emerging applications in semiconductor industry and biotechnology. Journal of Microelectromechanical Systems 15(4), 957–966 (2006)
6. Wixforth, A.: Acoustically driven planar microfluidics. Superlattices and Microstructures 33(5) (2004)
7. Tikka, A.C., Al-Sarawi, S., Abbott, D.: A Remotely Interrogatable Passive Microactuator using SAW Correlation. In: Proc. of 3rd International Conference on Sensing Technology, November 2008, pp. 46–51 (2008), ISBN: 978-1-4244-2176-3
8. Thiele, J.A., da Cunha, M.P.: High temperature saw gas sensor on langasite. In: Proc. of IEEE Sensors, October 2003, vol. 2, pp. 769–772 (2003)
9. Bastermeijer, J., Jakoby, B., Bossche, A., Vellekoop, M.J.: A novel readout system for microacoustic viscosity sensors. In: Proc. of 2002 IEEE Ultrasonics Symposium, October 2002, vol. 1, pp. 489–492 (2002)
10. Li, Y., Yang, M., Ling, M., Zhu, Y.: Surface acoustic wave humidity sensors based on poly(p-diethynylbenzene) and sodium polysulfonesulfonate. Sensors and Actuators. B 122(2), 560–563 (2007)
11. Luo, C.-P.: Detection of antibody-antigen reactions using surface acoustic wave and electrochemical immunosensors. PhD thesis, Ruperto-Carola University of Heidelberg (2004)
12. Sakong, J., Roh, Y., Roh, H.: 3f-2 saw sensor system with micro-fluidic channels to detect DNA molecules. In: Proc. of IEEE Ultrasonics Symposium, October 2006, pp. 548–551 (2006)
13. Mitsakakis, K., Tserepi, A., Gizeli, E.: Saw device integrated with microfluidics for array-type biosensing. Microelectronic Engineering 86(4) (2009)
14. Strobl, C.J., von Guttenberg, Z., Wixforth, A.: Nano- and pico-dispensing of fluids on planar substrates using saw. IEEE Trans. on Ultrasonics, Ferroelectrics, and Frequency Control 51(11), 1432–1436 (2004)
15. Nguyen, N.-T., White, R.M.: Acoustic streaming in micromachined flexural plate wave devices: numerical simulation and experimental verification. IEEE Trans. on Ultrasonics, Ferroelectrics, and Frequency Control 47(6), 1463–1471 (2000)
16. Rife, J.C., Bell, M.I., Horwitz, J.S., Kabler, R.C., Auyeung, R.C.Y., Kim, J.: Miniature valveless ultrasonic pumps and mixers. Sensors and Actuators. A 86(1) (2000)
17. Kondoh, J., Shimizu, N., Matsui, Y., Shiokawa, S.: Liquid heating effects by saw streaming on the piezoelectric substrate. IEEE Trans. on Ultrasonics, Ferroelectrics, and Frequency Control 52(10), 1881–1883 (2005)
18. Qi, Q., Brereton, G.J.: Mechanisms of removal of micron-sized particles by high-frequency ultrasonic waves. IEEE Trans. on Ultrasonics, Ferroelectrics, and Frequency Control 42(4), 619–629 (1995)
19. Changliang, X., Mengli, W.: Stability analysis of the rotor of ultrasonic motor driving fluid directly. Ultrasonics 43(7), 596–601 (2005)
20. Frampton, K.D., Minor, K., Martin, S.: Acoustic streaming in micro-scale cylindrical channels. Applied Acoustics 65(11), 1121–1129 (2004)
21. Yu, H., Kwon, J.W., Kim, E.S.: Microfluidic mixer and transporter based on pzt self-focusing acoustic transducers. IEEE Trans. on Ultrasonics, Ferroelectrics, and Frequency Control 15(4), 1015–1024 (2006)
22. Batra, R.C., Porfiri, M., Spinello, D.: Review of modeling electrostatically actuated microelectromechanical systems. Smart Materials and Structures 16(6), R23–R31 (2007)

23. Sounart, T.L., Michalske, T.A., Zavadil, K.R.: Frequency-dependent electrostatic actuation in microfluidic mems. Journal of Microelectromechanical Systems 14(1), 125–133 (2005)
24. Avdeev, I.V.: New formulation for finite element modeling electrostatically driven microelectromechanical systems. PhD thesis, University of Pittsburgh, ch. 3 (2003)
25. Moaveni, S.: Finite Element Analysis: Theory and Applications with ANSYS, 3rd edn. Prentice Hall, Englewood Cliffs (2007)
26. ANSYS Inc. ANSYS 9.0 Training Manual (2008)
27. Lai, Z.C.: Finite element analysis of electrostatic coupled systems using geometrically nonlinear mixed assumed stress finite elements. Master's thesis, University of Pretoria, ch. 4 (2007)
28. Galambos, P., Czaplewski, D., Givler, R., Pohl, K., Luck, D.L., Benavides, G., Jokiel, B.: Drop ejection utilizing sideways actuation of a mems piston. Sensors and Actuators. A 141(1), 182–191 (2008)
29. Kaajakari, V., Sathaye, A., Lal, A.: A frequency addressable ultrasonic microfluidic actuator array. In: Proc. of 11th International Conference on Solid State Sensors and Actuators Transducers 2001/Eurosensors XV, June 2001, pp. 958–961 (2001)

Flexible Piezoelectric Tactile Sensors with Structural Electrodes Array

Cheng-Hsin Chuang

Department of Mechanical Engineering,
Southern Taiwan University,
Tainan, Taiwan

Abstract. This study provides an efficient and feasible solution for enhancing the sensitivity of traditional piezoelectric tactile sensors based on introducing structural electrodes upon the sensing material. A sandwich structure for a flexible tactile sensor consists of top and bottom soft substrates made of Polystyrene; between these, a piezoelectric thin film, PVDF, and a PDMS microstructure array are utilized as sensing materials and microstructures, respectively. Experimental results showed that the output voltage was linear with a contact force from 10N to 0.5 N and good reliability within the low frequency range 1 ~ 100 Hz. In addition, shape recognition also can be achieved as the objective contacted using the 4 by 4 electrode array. The effects of structural electrodes on the enhancement of sensitivity were also numerically simulated by finite element method (FEM) and verified experimentally by dynamic measurement system, respectively. In general, the flexible tactile sensor developed in this study can be applied not only for detecting the degree of contact force but also as a human physiology monitoring system, measuring pulse, heart rate and blood pressure, etc.

1 Introduction

There are five main exteroceptive senses in humans, i.e. sight, hearing, taste, smell and touch. For consumer electronics, developing a better interface between people and technology has become an important driving force for innovation. Two human senses: sight and hearing, are now widely used in computer systems and hand-held devices for visual and audio communication (e.g. webcam, microphone, etc.). Recently, the sense of touch has drawn the attention of the manufacturers of hand-held devices. Touch is an intrinsic and primary means of perception for people interacting with the environment; for instance, touch panels have been built into mobile phones and many other devices. However, touch panels can usually sense only the location of touch on a display without providing other information such as the magnitude and direction of the contact force, and the shape and distribution of the touch area.. Nicholls and Lee [1] defined a tactile sensor as *a device or system that can measure a given property of an object or contact event through physical contact between the sensor and the object*. Tactile sensors, therefore, offer exciting possibilities in force-based interfaces measuring the degree of physical contact between a sensor and an object that can be applied to industrial and biomedical monitoring, inspection, and recognition as well as to consumer electronics. Because touch takes many forms and includes the detection of shape, texture, friction, force, pain, temperature and many other related physical properties, tactile sensing through the skin is difficult to imitate by a simple transduction of one physical property into an electronic signal. In the last two decades, micro-electro-mechanical systems (MEMS) have offered great potential in fabricating high-density sensor arrays and integrating multifunctional sensors on a single chip. A wide variety of MEMS tactile sensors have been demonstrated [2-13]; these tactile sensors can usually be classified into piezoresistive [2-7], capacitive [8, 9], piezoelectric [10-12] and optical [13], based upon the type of sensing mechanism.

Among these tactile sensors, the piezoresistive-type tactile sensor is the most commonly used and has been studied extensively due to its compatibility with silicon micromachining technology which allows high resolution of the sensory array to be easily achieved. However, some challenges, such as the hysteresis phenomenon and the temperature dependence of resistivity, require solutions. On the other hand, the capacitive-type tactile sensor exhibits very high sensitivity and is suitable for large-area sensing; nevertheless, the force range that can be sensed is usually insufficient since the capacitor formed by two electrode plates with a gap between could be saturated when the force acting on the capacitor is large enough to make two electrodes contact one another. For the optical-type tactile sensor, optical fibers are usually embedded in a flexible substrate with their light attenuations providing information regarding contact force and position. It therefore has a few advantages, such as low hysteresis, and not being influenced by external electromagnetic fields and humid environments. Because the light source and optical measurement systems are rather bulky and expensive however, the applications are limited. As for the piezoelectric-type tactile sensor, an additional piezoelectric material, such as lead zirconate titanate (PZT), is required to generate an electric potential in response to applied mechanical stress. Most piezoelectric materials can be divided in two main groups: crystals and ceramics. Although tactile sensing based on the piezoelectric effect is quite straightforward, both crystal and ceramic piezoelectric materials are difficult to attach to a curved surface such as robotic fingers. Hence, organic ferroelectrics such as poly(vinylidene fluoride) (PVDF), are commercially available and are regarded as promising materials that could be used as tactile sensors due to their particular characteristics [14-18]:

- High piezoelectric voltage sensitivity.
- Flexibility, thinness, and light-weight.
- Responsiveness over a wide frequency range.
- Ruggedness and inertness to chemical agents.

Basically, an ideal tactile sensor requires the following characteristics. First, a tactile sensor must be thin, flexible and able to cover a large area like the human skin. Second, a distributed array of detectors must be arranged with high density. Last of all, the sensor must be capable of withstanding external physical forces and chemical pollution. Yu *et al.* [18] demonstrated 8 × 8 distributed flexible tactile sensors by using PVDF film and flexible circuitry. However, the threshold of the sensor was about 0.7 N, which was insufficient for practical application; with limited sensitivity of 0.2 mV/N, additional signal amplification was needed. Consequently, both sensitivity and the threshold of contact force still have room for improvement in PVDF-based tactile sensors.

In order to enhance the signal output of PVDF-based tactile sensors, we introduced the concept of structural electrodes which mimic the rough surface of human skin upon the receptors, as shown in Fig. 1. Human skin is made up of multiple layers. The outmost layer is the epidermis, beneath lies the dermis layer and the layer closest to the bone is the subcutaneous layer. Under the epidermis, nipple-like structures called dermal papillae which cover the perceptional receptors can be found on the dermis surface. As an analogy to the nipple-like structure, we introduced the so-called "structural electrode" to replace the thin-film electrode in the traditional design shown in Fig. 2. Contrary to what occurs with a thin-film electrode in the traditional tactile sensor, the effect of stress concentration can be generated in the piezoelectric material while the contact force is transferred through the microstructures instead of uniform contact on the

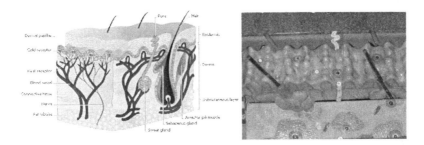

Fig. 1. Multilayer structure of human skin

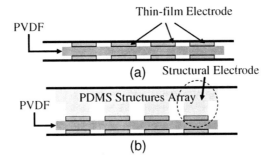

Fig. 2. Cross-sectional structure of (a) traditional PVDF-based tactile sensor with thin-film electrode array; (b) new type of PVDF-based tactile sensor with structural electrode created by adding a layer of PDMS microstructures array

piezoelectric material. The induced charge of piezoelectric material is therefore increased due to the stress concentrated underneath the microstructures. Although the structural electrode could enhance sensitivity, other side effects such as crosstalk effects, material selection of microstructure, size and shape effects of microstructure need to be considered. In the present paper, the effects of structural electrodes of different shapes and sizes were evaluated by finite element analysis (FEA). In addition, a 4 × 4 taxels flexible tactile sensor with PDMS microstructures array was fabricated and characterized by MEMS technology and dynamic measurement.

2 Simulation

A 3D model was established and analyzed by finite element software (ABAQUS) in order to investigate the influences of introducing microstructures, as shown in Fig.3. Three shapes and sizes for structural electrodes on PVDF film were examined by applying a given normal force. In acting as a normal force on the tactile sensor, the force is shared by the distributed structural electrodes array and directly transferred to the bottom PVDF film, assuming that the microstructure is a rigid column. If the total number of distributed structural electrodes in the contact force region remains constant in all kinds of simulation models, we know that each microstructure shares equal loading although they have different sizes and

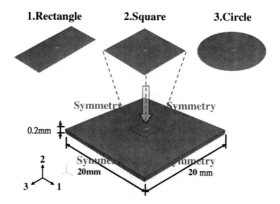

Fig. 3. 3D FEA Model for a single structural electrode: The PVDF film is 200 μm thick and 20 mm in square, the size of electrode is 9, 16, 25 mm² with different shapes

Table 1. Material properties of PVDF film for simulation

Density (kg/cc)	Young's modulus (GPa)	Poisson ratio	Dielectric constant (Farad/m)	d_{211}, d_{233} (m/Volt)	d_{222} (m/Volt)
1780	3	0.35	11×10^{-10}	23×10^{-12}	-33×10^{-12}

shapes. To evaluate these effects in the simulation, the identical loading for each kind of structural electrode was set as 500N. In addition, the PVDF film is 200 μm in thickness and its material properties and piezoelectric coefficients used in the simulation are listed in Table 1.

As the size of the structure electrode was 16 mm^2, the distribution of electrical potential on the surface of different shapes of PVDF film are illustrated in Fig. 4. The results indicate there is a plateau region of electrical potential corresponding to the compressive stress area underneath the structural electrode; the circular structural electrode possessed slightly higher electrical potential than the square and rectangular

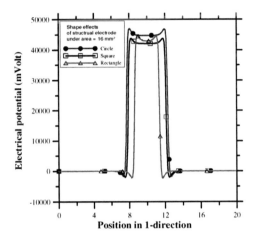

Fig. 4. The electrical potential profile across the center of the contact area and parallel in 1-direction, the area of the electrode is 16 mm^2 but electrode shapes vary

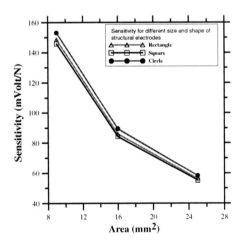

Fig. 5. The effects of size and shape on the sensitivity of tactile sensors with structural electrodes

electrodes. In addition, a peak value of electrical potential occurred at the edge of the structural electrode due to the stress concentration at the edge and the deformation of the PVDF film. Regarding the effects based upon the size of structural electrodes, sensitivity, defined as the ratio of output voltage (mV) to the applied force (N), was higher as the area of structural electrodes decreased, as shown in Fig. 5. Thus, a smaller structural electrode could effectively enhance the sensitivity of piezoelectric tactile sensors; however, the influence of electrode shape remains less significant than the size of the electrode.

In general, the cross-talk effect needs to be avoided as we wish to exactly identify the image of the contact area when an object or a force contacts a tactile sensor. Cross-talk means there is no contact at the taxel (tactile element) but the taxel still can gather the output signal due to the influence of other neighboring taxels. As shown in Fig. 4, a positive peak value of electrical potential occurred at the edge of structural electrodes and an inverse electrical potential also appeared near the structural electrode. This declined to zero when the distance from the edge of the structural electrode increased. This phenomenon can be attributed to the stress concentration at the edge of the mcirostructure and the elastic deformation of the substrate (PVDF) as shown in Fig. 6. Therefore, cross-talk effects could be found if the spacing between two structural electrodes is insufficient. Here, we defined the line distance of inverse electrical potential from the edge of a structural electrode to the zero potential as the influence range of cross-talk illustrated in Fig. 6. Then, the influence range can be normalized by the width of structural electrodes for all cases with different sizes and shapes as listed in Table 2. According to Table 2, when the spacing between each taxel is larger than the ratio to the electrode width, the cross-talk effect can be avoided. Obviously, the cross-talk effect is more serious with a smaller structural electrode and a circular structural electrode decreases the range of cross-talk effects. For example, the ratio of spacing to width for a 16 mm^2 taxel should be at least larger than 1 for all kinds of structural electrodes.

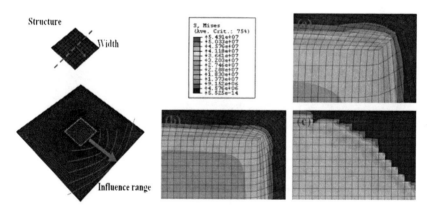

Fig. 6. The stress distribution at the edge of structural electrodes: (a) Rectagular; (b) Square; (c) Circle

Table 2. Cross-talk effect for different shapes and sizes of structural electrodes

Taxel Area (mm²)	Influence_Range Width_of_Electrode		
	Rectangle	*Square*	*Circle*
9	1.38	0.90	0.78
16	0.98	0.67	0.55
25	0.76	0.51	0.43

3 Tactile Sensor Design and Fabrication

The structure of the tactile sensor consisted of two parts. The bottom layer was the sensing material PVDF film (Measurement Specialties Inc.) patterned with distributed metal (Cr) electrodes by E-beam evaporation, photolithography and wet etching as shown in Fig. 7(a). The thicknesses of PVDF film and electrode layer were 52μm and 1500Å, respectively. The top layer was the PDMS microstructures array made by the molding process with a precision machining steel mold, and the pitch, diameter and height of PDMS microstructure, as shown in Fig. 7(c), were of the same value (2 mm). Finally, the top and bottom layers were bonded together and packaged with two polyester films used as electrical insulation.

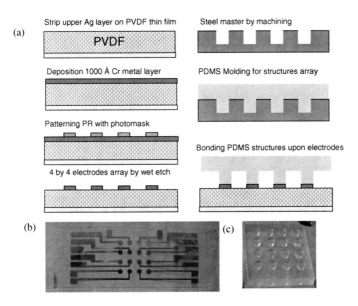

Fig. 7. The process flow for the fabrication of a 4 × 4 distributed flexible tactile sensor with PDMS microstructures array, the thickness of PVDF film is 52 μm, the radius of each structural electrode is 1 mm and the center-to-center spacing is 4 mm, the total size of sensor is 2 cm × 2cm. (b) The distributed electrode and (c) the resulting structure of PDMS.

4 Experimental Setup

For a piezoelectric tactile sensor, it is difficult to measure a static force due to induced charges by a static force that could only be detected as the force is loading. The output voltage will decline rapidly in a very short time because, there are no further charges that can be induced from piezoelectric film after the contact force in a static state. Therefore, a dynamic test system was built to obtain the output characteristics of a tactile sensor. The applied force was excited by a shaker (Data Physics Corp.) and calibrated by a force sensor (PCB209C02, PCB Piezotronics, Inc.). The signals of 16 taxels were scanned sequentially by an analog multiplexer (MAX406, 16 channels, Maxim Integrated Products, Inc.) and amplified by a homemade charge amplifier with a 60 Hz notch filter. The signal processing for

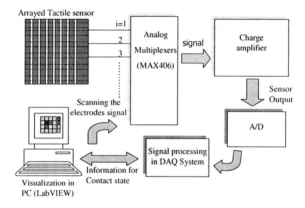

Fig. 8. Block diagram of signal processing flow

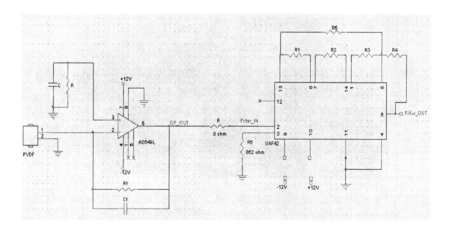

Fig. 9. Charge amplifier with a 60Hz notch filter

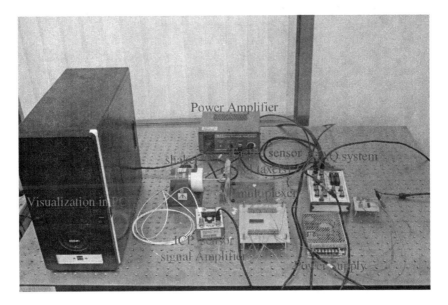

Fig. 10. Configuration of experimental setup

force sensors and tactile sensors was controlled by PC-based LabVIEW programming. The signal process flow and the circuit design of the charge amplifier with a notch filter are shown in Fig. 8 and Fig. 9, respectively. The shaker generated a horizontal periodic force which was transferred by a shaft mounted with a piece of steel plate upon the tactile sensor for a uniform loading condition as illustrated in the Fig. 10. The processed signals of sensor output were visualized on a personal computer; the shape and force distribution of the contact object were also obtained in real time.

5 Experimental Results

5.1 Output Characteristics of Sensor

First, we characterized the tactile sensor by a sinusoidal force of 1.5 N under 2 Hz as shown in Fig. 11(b). By comparison with the signal of the force sensor mounted on the shaft and shown in Fig. 11(a), the tactile sensor output showed the time delay of 50 ms due to signal processing and slight noise form the AC power source of the amplifier at 60 Hz. In addition, the dynamic response of the PVDF sensor exhibited high stability and low hysteresis; however, a DC bias of 20 mV was observed in the experimental results. The voltage offset can be minimized by impressing a pre-charge DC bias as proposed by Edward S. Kolesar [10]. In our experiments, we extracted the absolute value of the output signal amplitude by using a LabVIEW program for the real-time monitoring and shape recognition; hence we didn't impress any pre-charge DC bias on our tactile sensor.

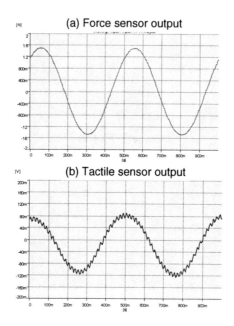

Fig. 11. Dynamic responses of force snesor and tactile sensor under applied a sinusoidal force of ±1.5 N under 2 Hz

The relationship between the applied forces and the output voltages of a tactile sensor without a charge amplifier is shown in Fig. 12. The threshold of the sensor is about 0.5 N and it has good linearity under 4 N. Fig. 13 shows the frequency response of the tactile sensor; the horizontal axis is the frequency of an applied 1

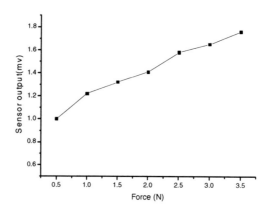

Fig. 12. Calibration of tactile sensor outputs

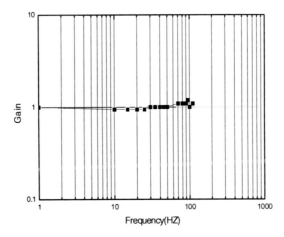

Fig. 13. Frequency response of tactile sensor

N force from 1 to 120 Hz; the vertical axis is the gain of the sensor output compared to the value at 1 Hz. As for the results, the bandwidth of linear response is 50 Hz, and the voltage fluctuation can be observed in the high frequency range (80 to 120 Hz). Hence, an additional compensation is needed for measuring a high frequency response.

5.2 Sensitivity Enhancement by Structual Electrodes

According to the simulation results, the sensitivity can be enhanced by adding a structural layer upon the sensing electrodes. For quantifying the difference observed with and without PDMS structures, a uniform loading on a specific contact area is necessary; therefore, a sinusoidal force of 1 N under 2 Hz was applied through a piece of steel plate placed in the central four taxels. The output voltages for each taxel can be read out by the LabVIEW program and visualized by MATLAB software as listed in Table 3 and shown in Fig. 14, respectively. The sensitivity of a

Table 3. Sensitivity for tactile sensors with thin-film electrodes and structural electrodes

Taxel	Sensitivity (mV/N)		Enhanced percentage (%)
	A: Without PDMS Structures	B: With PDMS Structures	(B-A)/A
(2,2)	4.74	5.80	22
(2,3)	4.26	5.41	27
(3,2)	4.81	6.28	30
(33)	4.53	7.32	62
Average	4.58	6.20	35

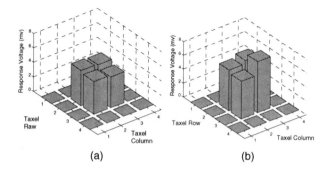

Fig. 14. 3D images of the output voltages as uniformly applied 1 N external force on the central four taxels: (a) without PDMS structures; (b) With PDMS structures

tactile sensor with PDMS structures can be enhanced on average by 35% as compared to using a conventional sensor with thin-film electrodes. However, variation between each taxel is increased as the uniform loading is transferred by the PDMS structures array. This can be attributed to fabrication errors in PDMS structures and misalignment between microstructures and electrodes.

5.2 Shape Recognition Results

A human-machine interface was established by the LabVIEW program as shown in Fig. 15. The signals of force sensors and tactile sensors can be graphically represented on the screen. Furthermore, a color 2D image can be also displayed on the intensity graph area after the output voltage of each taxel is scanned sequentially by a multiplexor and the amplitude extracted by a fast Fourier transform (FFT) algorithm.

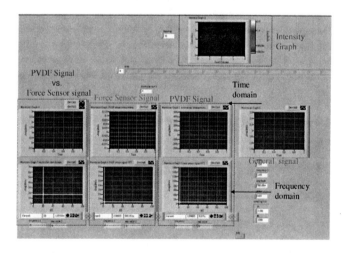

Fig. 15. Human-machine interface constructed by the LabVIEW program

Two pieces of steel plate of different shapes were mounted on the shaker, in contact with the tactile sensor for shape recognition. As the results in Fig. 16 show, the output voltages of contact taxels were 100 times higher than the other noncontact taxels. Consequently, the shape of an object can be easily identified based on this distributed flexible tactile sensor.

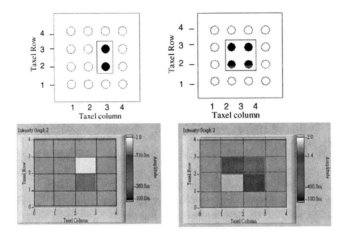

Fig. 16. Two different object shapes can be identified by 2D intensity images

6 Conclusion

In this study, we introduced the concept of using a structural electrode in the PVDF-based tactile sensor for the enhancement of sensitivity. A 4 by 4 distributed flexible tactile sensor was designed and fabricated based on the MEMS technology and Molding process. The experimental results show that the sensitivity can be increased roughly 30% due to microstructural effects; however, the contact conditions between PDMS structures and metal electrodes have to be improved for ensuring the accuracy of force detection. The human-machine interface based on LabVIEW programming is able to achieve the function of shape recognition and force measurement in real time. In addition to sensitivity enhancement, more advantages in the detection of incident slippage and multi-directional forces have been investigated; we will present our work in the following publication.

References

[1] Lee, M.H., Nicholls, H.R.: Tactile sensing for mechatronics-a state of the art survey. Mechatronics 9, 1–31 (1999)
[2] Kane, B.J., Cutkosky, M.R., Kovacs, G.T.A.: A Traction Stress Sensor Array for Use in High-Resolution Robotic Tactile Imaging. Journal of Microelectromechanical Systems 9(2), 425–434 (2000)

[3] Hasegawa, Y., Shikida, M., Shimizu, T., Miyaji, T., Sasaki, H., Sato, K., Itoigawa, K.: Amicromachined active tactile sensor for hardness detection. Sensors and Actuators A 114, 141–146 (2004)
[4] Engel, J., Chen, J., Fan, Z., Liu, C.: Polymer micromachined multimodal tactile sensors. Sensors and Actuators A 117, 50–61 (2005)
[5] Noda, K., Hoshino, K., Matsumoto, K., Shimoyama, I.: A shear stress sensor for tactile sensing with the piezoresistive cantilever standing in elastic material. Sensors and Actuators A 127, 295–301 (2006)
[6] Dao, D.V., Sugiyama, S.: Development of 4-DOF Soft-Contact Tactile Sensor and Application to Gripping Operation of Robotics Fingers. In: Proceeding of IEEE SENSORS 2006, pp. 1321–1324 (2006)
[7] Ádáma, M., Mohácsy, T., Jónás, P., Dücső, C., Vázsonyi, É., Bársony, I.: CMOS integrated tactile sensor array by porous Si bulk micromachining. Sensors and Actuators A 142, 192–195 (2008)
[8] Ko, C.-T., Tseng, S.-H., Lu, M.S.-C.: A CMOS Micromachined Capacitive Tactile Sensor With High-Frequency Output. Journal of Microelectromechanical Systems 15(6), 1708–1714 (2006)
[9] Lee, H.-K., Chung, J., Chang, S.-I., Yoon, E.: Normal and Shear Force Measurement Using a Flexible Polymer Tactile Sensor With Embedded Multiple Capacitors. Journal of Microelectromechanical Systems 17(4), 934–942 (2008)
[10] Kolesar Jr., E.S., Dyson, C.S.: Object Imaging with a Piezoelectric Robotic Tactile Sensor. Journal of Microelectromechanical Systems 4(2), 87–96 (1995)
[11] Takashima, K., Horie, S., Mukaia, T., Ishida, K., Matsushige, K.: Piezoelectric properties of vinylidene fluoride oligomer for use in medical tactile sensor applications. Sensors and Actuators A 144, 90–96 (2008)
[12] Qasaimeh, M.A., Sokhanvar, S., Dargahi, J., Kahrizi, M.: PVDF-Based Microfabricated Tactile Sensor for Minimally Invasive Surgery. Journal of Microelectromechanical Systems 18(1), 195–207 (2009)
[13] Heo, J.-S., Kim, J.-Y., Lee, J.-J.: Tactile Sensors using the Distributed Optical Fiber Sensors. In: Proceeding of IEEE the 3rd International Conference on Sensing Technology, pp. 486–490
[14] Dargahi, J.: A piezoelectric tactile sensor with three sensing elements for robotic, endoscopic and prosthetic applications. Sensors and Actuators 80, 23–30 (2000)
[15] Yamada, Y., Maeno, T., Fujimoto, I., Morizono, T., Umetani, Y.: Identification of Incipient Slip Phenomena Based on The Circuit Output Signals of PVDF Film Strips Embedded in Artificial Finger Ridges. In: Proc. SICE Annual Conf. 2002, pp. 3272–3277 (2002)
[16] Dargahi, J., Najarian, S., Liu, B.: Sensitivity analysis of a novel tactile probe for measurement of tissue softness with applications in biomedical robotics. Journal of Materials Processing Technology 183, 176–182 (2007)
[17] Dargahi, J., Sedaghati, R., Singh, H., Najarian, S.: Modeling and testing of an endoscopic piezoelectric-based tactile sensor. Mechatronics 17, 462–467 (2007)
[18] Yu, K.-H., Yoon, M.-J., Kwon, T.-K., Lee, S.-C.: Distributed flexible tactile sensor system. International Journal of Applied Electromagnetics and Mechanics 18, 53–65 (2003)

A New Method for Direct Gravity Estimation and Compensation in Gyro-Based and Gyro-Free INS Applications

Ehad Akeila, Zoran Salcic, and Akshya Swain

Department of Electrical and Computer Engineering
University of Auckland,
Auckland, New Zealand

Abstract. This paper proposes a new method for estimating the changes of gravity and compensating its effects on acceleration measurements in Inertial Navigation System (INS) based applications. Gravity is a critical factor and its effects are usually included in the measurements obtained from accelerometers. This results in a need to eliminate gravity effects from the measurements in order to accurately determine the true accelerations of moving objects. The proposed method utilizes the measurements from the existing INS sensors without employing any additional sensors. This significantly reduces the complexity of INS based system design. Two different types of models have been developed based on the proposed gravity compensation method and they are applied on Gyro-Based and Gyro-Free INS based applications. The performance of both these models has been analyzed by simulations and experiments. Error analysis of the new method has been studied and related equations have been derived. Results of simulations demonstrate that by including the proposed gravity compensation model, the errors in the acceleration measurements reduces in the Gyro-Free INS applications.

1 Introduction

Inertial Navigation Systems (INS) have been utilized in many applications which involve tracking of moving objects. The tracking process essentially requires measurement of both linear accelerations and rotational motions. These motions are estimated in the INS systems using either one of the two commonly known methods; (a) Gyro-Based INS (GBINS) or (b) Gyro-Free INS (GFINS). In the GBINS method, accelerometers and gyroscopes are employed in measuring the linear accelerations and the angular rotations, respectively. This method is the common INS tracking technique used in a wide range of applications. On the other hand, the GFINS method does not involve the use of gyroscopes for measuring rotational motions; it fully relies on accelerometers in measuring both the linear and rotational motions. This is achieved by placing the accelerometers in a certain configuration, such as sides of a cube [1], in order to attain a feasible GFINS system [2-4]. The applications that use this method are limited as this technique is still under research.

In both GBINS and GFINS methods, accelerometers play important role; the performance of the INS systems significantly depends on measurements acquired from these sensors. These sensors measure the acceleration forces along their sensitive axes. The applied forces may be one of two types (a) *dynamic*, caused by moving or vibrating the accelerometers (this is related to linear acceleration measurement), or they could be (b) *static* like the constant force of gravity applied on the sensor (tilt angle measurement). Accelerometers have best performance when they are exposed to either one of the two types of forces. When applying a static force, like gravity, accelerometers are capable of measuring the tilt angle of objects with respect to the earth gravitational direction. This property has been valuable for a number of applications, such as those in the biomedical area [5, 6]. In these applications, accelerometers are employed to measure the tilt angle of parts of the human body in order to estimate the workload exerted on these parts and find any muscular disorders. In such applications the moving parts are expected to have one type of motion; static acceleration in this case. However, a critical limitation comes into picture when objects move randomly or have combined motions, consisting of linear acceleration and rotations (which cause changes in tilt angles). In this case, the outputs of accelerometers will be the sum of effects of these motions and it becomes difficult to distinguish which type of motion has been applied [7, 8]. This combination, if not properly compensated, often results in large errors.

Several techniques have been proposed in the past to eliminate the effects of gravity from the outputs of accelerometers. Most of these techniques rely on models that can predict the changes of gravity in certain type of motions [7, 9, 10]. In [9] the gravitational values corresponding to the coordinates of some regional areas are pre-stored on an external memory. Based on the initial coordinates estimated from the INS system, the gravity values are extracted from the memory and then used to compensate the INS measurements. Other gravity compensation techniques depend on employing additional sensors, which are external to the main INS system, to supply gravity information as in [11] and [12]. In other applications as in [13], the gravity is compensated using a low pass filter, which filters out the gravitational changes which normally have low frequencies in the case of small autonomous helicopters. However, in applications where motions of the object are random, such as the smart pebble [14], it is difficult to accurately predict and estimate the changes in gravity using external sensors due to the limitations of size of the overall system and the complexity of its motions.

This paper proposes a new method for accurate measurement and compensation of gravitational changes directly from the measurements acquired from the sensors included in the INS system, without using any additional sensors or predefined models. This in turn reduces the complexity of the INS based design as well as the overall cost. Based on the proposed method, models have been developed both for Gyro-Based and Gyro-Free INS applications. The performance of the proposed gravity compensation techniques has been illustrated both via simulations and experiments. The effect of the new method has been studied through derivation of error equations for each INS technique. It has also been shown that the new method results in reduction of the errors in GFINS models.

The rest of the paper is organized as follows: Section 2 gives basic equations of the output of the accelerometer. Section 3 describes the fundamental equations and models used in INS tracking systems. The new gravity compensation algorithms are described in section 4 followed by validation of models using simulation results in section 5. Error analysis and error equations are described in section 6. Section 7 shows the experimental results followed by conclusion in section 8.

2 Accelerometer Output

In INS applications, accelerometers are designed to be mounted on rigid bodies to measure the force per unit mass along their sensing direction. This force is called *specific force* $\hat{F}$ [15]. The vector sum of $\hat{F}$ and the gravity g is the total inertial acceleration of the body, that is, $\hat{F} + g = ma$, $m = 1$. Therefore, the specific force is $\hat{F} = a - g$. The accelerometer is designed to measure this force magnitude projected along the sensing axis (or orientation). Let θ_I represents the orientation of the sensor and r_I represents the location of the sensor relative to a fixed inertial frame. The output of the accelerometer is given by:

$$A = \langle a - g, \theta_I \rangle = \langle \ddot{r}_I - g, \theta_I \rangle \tag{1}$$

In practice, accelerometers measure two types of accelerations, static acceleration (gravity) and dynamic accelerations. If a sensor is mounted on an object having acceleration a, then the physical output of the accelerometer will be given by:

$$V_{out} = K_a a + K_g g \tag{2}$$

where V_{out} is the output voltage of the accelerometer, K_a is a component of the actual applied acceleration vector (a), and K_g is a component of the actual applied gravity vector.

In applications which need tilt measurements, accelerometers are one of the best devices which are being used for measuring tilt angles. In such applications, the dynamic acceleration will be equal to zero, so the output of the accelerometer is only gravity according to (2). However, in applications where the object has dynamic acceleration combined with change in the tilt angle, it becomes difficult to measure linear accelerations due to the problems in separating tilt from acceleration measurements.

3 INS Tracking Models

This section describes the essential models used for tracking objects using both Gyro-Based and Gyro-Free INS methods.

A. Gyro-Based INS Models

These are the commonly used models in INS based tracking applications. They use three accelerometers in addition to three gyroscopes in order to measure the acceleration of objects in three directions (x, y and z). The sensors are normally placed orthogonally to each other as shown in Figure 1. The accelerometers measure linear accelerations, while gyroscopes measure the rotations of the object. Once the measurements from the sensors are obtained, the Euler angle transformation is applied to calculate the accelerations with respect to the initial orientation of the object using [8]:

$$\begin{bmatrix} Ax_r \\ Ay_r \\ Az_r \end{bmatrix} = C_b^r \bullet \begin{bmatrix} Ax_b \\ Ay_b \\ Az_b \end{bmatrix} \tag{3}$$

where Ax_r, Ay_r, Az_r are the accelerations with respect to the initial orientation (reference frame) of the body, Ax_b, Ay_b and Az_b, are the measured accelerations along each axis of the object (moving frame) and C_b^r is the rotation matrix which transforms the measured accelerations from the body frame to the reference frame. It is described as:

$$C_b^n = \begin{bmatrix} \cos\theta\cos\psi & -\cos\phi\sin\psi+\sin\phi\sin\theta\cos\psi & \sin\phi\sin\psi+\cos\phi\sin\theta\cos\psi \\ \cos\theta\sin\psi & \cos\phi\cos\psi+\sin\phi\sin\theta\cos\psi & -\sin\phi\cos\psi+\cos\phi\sin\theta\sin\psi \\ -\sin\theta & \sin\phi\cos\theta & \cos\phi\cos\theta \end{bmatrix} \tag{4}$$

where ϕ, θ and φ are the rotational angles around the x, y and z directions respectively. Note that equation (4) is valid if Ax_b, Ay_b and Az_b are the true linear

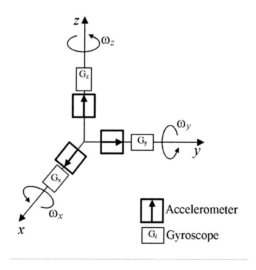

Fig. 1. Euler angle transformation

accelerations along each axis. However, when accelerometers are used in measuring those accelerations, the gravitational component need to be filtered out to obtain the true acceleration values.

B. Gyro-Free INS Models

The Gyro-Free INS model is derived based on the conventional six-accelerometer cubic configuration as shown in Figure 2 [15]. Each sensor has a location of u and orientation of θ. The outputs of the accelerometers are modelled according to the following equation:

$$A_{(1..6)} = \langle F^T(\ddot{R}-g), \theta_i \rangle + \langle (\dot{\Omega}+\Omega^2)u, \theta_i \rangle \tag{5}$$

where F^T is the rotation matrix, $\ddot{R}$ is the final 3D linear acceleration of the cube, Ω is a 3x3 matrix which models the 3D angular velocity of the cube, and is defined as:

$$\Omega = \begin{bmatrix} 0 & -\omega_z & \omega_y \\ \omega_z & 0 & -\omega_x \\ -\omega_y & \omega_x & 0 \end{bmatrix} \tag{6}$$

Note that the first term in (5), $\langle F^T(\ddot{R}-g), \theta_i \rangle$, computes the linear acceleration while the second term $\langle (\dot{\Omega}+\Omega^2)u, \theta_i \rangle$ computes the rotational motion.

Based on the above equations, the linear and rotational motions are described by the matrices L and $\dot{\omega}$ respectively

$$\dot{\omega} = \begin{bmatrix} \dot{\omega}_x \\ \dot{\omega}_y \\ \dot{\omega}_z \end{bmatrix} = \frac{1}{2\sqrt{2}l} \begin{bmatrix} A_1 - A_2 + A_5 - A_6 \\ -A_1 + A_3 - A_4 - A_6 \\ A_2 - A_3 - A_4 + A_5 \end{bmatrix} \tag{7}$$

$$L = \frac{1}{2\sqrt{2}} \begin{bmatrix} A_1 + A_2 - A_5 - A_6 \\ A_1 + A_3 - A_4 + A_6 \\ A_2 + A_3 + A_4 + A_5 \end{bmatrix} + l \begin{bmatrix} \omega_2 \omega_3 \\ \omega_1 \omega_3 \\ \omega_1 \omega_2 \end{bmatrix} \tag{8}$$

Note that the matrix L also equals to [15]:

$$L = F^T(\ddot{R}-g) \tag{9}$$

In this equation, gravity is compensated from the system measurements; however it is assumed that gravitational models are available and supplied to the system from external sources. In section 4 we show how gravity can directly be estimated from the GFINS cube accelerometers.

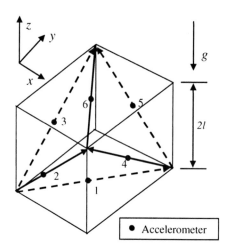

Fig. 2. The six-accelerometer cube configuration

4   Gravity Compensation Models

The gravity compensation algorithm developed in this work is based on the assumption that when the object is left stationary, the accelerometers measure only the gravity. At stationary conditions where the linear acceleration $a = 0$ equations (1) and (2) reduces to

$$A = \langle 0 - g, \theta_i \rangle = \langle -g, \theta_i \rangle \tag{10}$$

$$V_{out} = K_g g \tag{11}$$

Figure 3 shows the flowchart of the developed algorithm [16]. The new developed method suggests that the object has to be left stationary for few seconds (initialization phase) before the start of the tracking process from the INS device at ($t < 0$). By doing so, the initial pure gravitational values can be measured and then stored in memory. When the object starts moving at ($t > 0$), real-time gravitational outputs can be measured based on the values stored in the memory as described in the following sections. The real-time gravity values, or changes in the tilt angle, are obtained by multiplying initial stored values by the rotational matrix which is estimated within the INS system. Thus, gravity can be compensated from the accelerometer outputs by subtracting those gravity measurements from the outputs of the accelerometers while the object is subjected to motion. The following sections describe the algorithms and models of measuring gravity directly from the sensors in Gyro-Based and Gyro-Free INS applications in real-time.

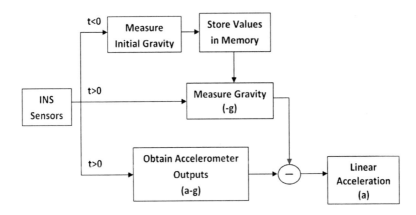

Fig. 3. Gravity compensation algorithm

A. Compensation in Gyro-Based Methods

Measurement of gravity is carried out by measuring the tilt angle of the accelerometer and then multiplying it with the acceleration value (9.81 m/s^2). Because the rotational matrix C_b^r described in (4) can transform measurements from the moving frame to the initial reference frame, the real-time changes in tilt angles in the moving frame (gravity values) can be measured in this type of INS applications at (t>0) by multiplying the initial gravitation values in the reference (tilt values obtained at t<0) which are stored in the memory by rotational matrix C_b^r using

$$\begin{bmatrix} g_x \\ g_y \\ g_z \end{bmatrix}_{t>0} = C_b^r \begin{bmatrix} -g_x \\ -g_y \\ -g_z \end{bmatrix}_{t<0} \tag{12}$$

Thus, equation (3) can be modified to measure the linear accelerations directly from the accelerometers (Ax$_s$ Ay$_s$ and Az$_s$) after eliminating the gravity factor from their outputs using (12). This is done as follows:

$$\begin{bmatrix} Ax_r \\ Ay_r \\ Az_r \end{bmatrix} = C_b^r \bullet \left(\begin{bmatrix} Ax_b - g_x \\ Ay_b - g_y \\ Az_b - g_z \end{bmatrix} + \begin{bmatrix} g_x \\ g_y \\ g_z \end{bmatrix}_{t>0} \right) \tag{13}$$

From (2), the output of the accelerometer is (a-g). Thus

$$\begin{bmatrix} Ax_r \\ Ay_r \\ Az_r \end{bmatrix} = C_b^r \bullet \left(\begin{bmatrix} Ax_s \\ Ay_s \\ Az_s \end{bmatrix} + \begin{bmatrix} g_x \\ g_y \\ g_z \end{bmatrix}_{t>0} \right) \tag{14}$$

By substituting the gravity components of (12) into (14) gives

$$\begin{bmatrix} Ax_r \\ Ay_r \\ Az_r \end{bmatrix} = C_b^r \bullet \begin{bmatrix} Ax_s \\ Ay_s \\ Az_s \end{bmatrix} + \left(C_b^r\right)^2 \bullet \begin{bmatrix} g_x \\ g_y \\ g_z \end{bmatrix}_{t<0} \qquad (15)$$

B. Compensation in Gyro-Free INS Methods

Under stationary conditions both the linear acceleration $\ddot{R}$ and the matrix Ω (described in sec 3.2) will be equal to zero. Because $\dot{F} = F\Omega$ [15], the matrix L described in (9) can be rewritten as

$$L_{t<0} = F^T(0-g) = I(-g_{t<0}) = -g_{t<0} \qquad (16)$$

Therefore, the real-time gravity measurement can be obtained by multiplying the initial gravity or L values (at t<0) by rotational matrix F

$$g_{t>0} = -FL_{t<0} \qquad (17)$$

This measurement can be substituted in the original L matrix in (9) at t >0 to calculate the real acceleration which becomes

$$\ddot{R} = FL_{t>0} - FL_{t<0} = F(L_{t>0} - L_{t<0}) \qquad (18)$$

5 Verification Using Simulation Models

The developed gravity compensation models have been created in MATLAB for validation purposes. The effectiveness of the model was illustrated by applying a type of motion which only produces changes in gravitational output without creating any linear acceleration. Hence, the success of each method is judged based on completely filtering out the gravity and making the system nonresponsive to the gravity changes in case of both Gyro-Based and Gyro-Free methods. To test the compensation method, the INS models, GBINS and GFINS, were applied to an oscillating rotation angles of 180° around x-axis as shown in Figure 4. This type of motion is applied in order to make sure that the rotational motion only changes gravity without making any linear accelerations. This is achieved as the radius of the motion is zero and since ($a_{linear}=\omega^2.r$), where ω is the rotational velocity applied, then a_{linear} will be also zero. In this motion, the changes in gravity will be observed in the y and z axes. Figure 5 and Figure 6 show the simulation results of verifying the algorithm in the GBINS and GFINS respectively. From the figures, it is obvious that the method can efficiently compensate for gravity. By comparing the gravity estimation errors shown in figures 5(b) and 6(b), it is clear that the performance in the GBINS is better than in the GFINS. This is mainly due to the higher amount of computations and approximations required in the simulating the GFINS systems. However, the performance becomes different when testing the actual hardware models of these systems as will be discussed in section 8.

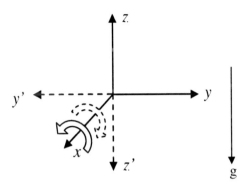

Fig. 4. A schematic showing the oscillating motion applied on the GFINS and GBINS systems to simulate the effect of gravity on these systems

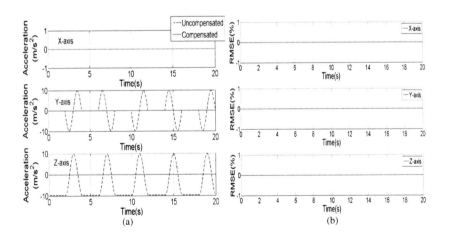

Fig. 5. Simulation results verifying the gravity compensation algorithm applied on the Gyro-Based INS method. (a) The compensated and uncompensated gravity output. (b) The error percentage in estimating the actual gravitational values.(RMSE: Root Mean Square Error)

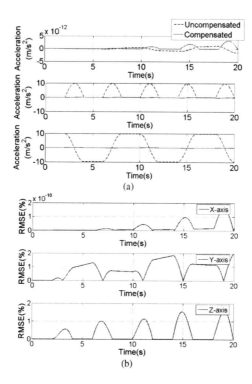

Fig. 6. Simulation results verifying the gravity compensation algorithm applied on the Gyro-Free INS method. (a) The compensated and uncompensated gravity output. (b) The error percentage in estimating the actual gravitational values.

6 Error Analysis of the New Compensation Method

The following sections describe the error analysis of the new gravity compensation method when it is applied to both GBINS and GFINS models.

A. Error Analysis of the GBINS

The error equations derived in this section are based on (15) which shows the final GBINS equation after applying the new gravity compensation algorithm. The equation can be rewritten as follows:

$$Ar_{ideal} = C_b^r \bullet As + \left(C_b^r\right)^2 \bullet g_{t<0} \qquad (19)$$

where $Ar = [Ax\ Ay\ Az]^T$. From (19), it is seen that the accuracy of acceleration estimation depends on the accuracy of the rotational matrix C_b^r and the accelerometers measurements As. Thus, there exist essentially two main sources of error such as

- $e1$: is the error due to gyroscopes which is reflected in computing the rotation matrix C_b^r

- $e2$: is the measurement error from the accelerometers which affects As as well as the initial gravity $g_{t<0}$

In corporating the effects of these errors in to the model of (19) gives

$$Ar_{error} = (C_b^r + e1) \bullet (As + e2) + (C_b^r + e1)^2 \bullet (g_{t<0} + e2) \qquad (20)$$

By expanding (20), the final error equation becomes

$$Ar_{error} = Ar_{ideal} + e1^2(g_{t<0} + e2) + e1\left[(1+2C_b^r)e2 + As + 2C_b^r \cdot g_{t<0}\right] + e2\left[C_b^r + (C_b^r)^2\right] \qquad (21)$$

From (21) it can be seen that any small errors in the accelerometers or gyroscopes could result in higher errors in measuring accelerations using GBINS method when the new gravity compensation algorithm is applied. Moreover, the new algorithm is more sensitive to gyroscopes errors ($e1$) as (21) contains more second order terms than the accelerometers errors ($e2$).

B. Error Analysis of the GFINS

Obtaining position of objects using equations of the GFINS cubic configuration requires more computational efforts compared to the equations used in the GBINS method. Consequently, the GFINS systems have more sources of errors compared to the GBINS systems. The location of the sensors and their orientations in the GFINS cube are critical and significantly affect the accuracy of the system.

To analyse the effect of the new gravity algorithm, consider equation (18) and add the term $e3$ as the total errors in estimating the $L_{t>0}$ values and the term $e4$ to represent the error in measuring the initial values $L_{t<0}$. Hence the error equation becomes as follows:

$$\ddot{R}_{error} = F(L_{t>0} + e3 - L_{t<0} - e4)$$
$$= \ddot{R}_{ideal} + F(e3 - e4) \qquad (22)$$

The above equation clearly shows that the new algorithm has a positive impact on reducing the GFINS errors. To confirm this effect, the GFINS simulation models were tested by assuming that there is 1% error in the location (position) of one of the sensors on the cube.

The model has been exposed to a rotational motion around circle of radius 10cm with angular acceleration of 0.1 rad/s^2. The simulation was performed under two conditions; (a) assuming ideal gravity measurements are supplied from external sources and (b) assuming that gravity is measured from the new gravity method. The error in estimating the final position travelled is plotted in Figure 7. From the figure, it is observed that the new gravity estimation method has reduced the error accumulation in calculating the position of the cube.

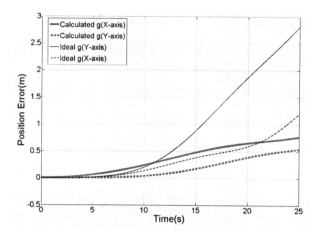

Fig. 7. Accumulation error in calculating the position of the cube when the gravity is measured using ideal values and using the new model

7 Experimental Results

Beside the verification of the new algorithm using simulation, it was necessary to verify it further by applying it on actual hardware models of the GBINS and GFINS. A similar type of motion as discussed in section 5 has been applied to each hardware prototype.

Considering the GBINS methods, the algorithm has been applied on two systems:

- Smart Pebble [14]: The smart pebble is based on low cost accelerometers and gyroscopes and used for monitoring the sediment transport inside river-beds. The gravitational changes were applied around the z-axis in this application. Figure 8 shows the recent developed prototype of the Smart Pebble as well as the output results.

- MAG 3D INS system: This is a complete INS sensors system in a small 3x3cm package. It contains orthogonal tri-axial accelerometer, gyroscopes and magnetic sensors [17]. This system has been particularly selected due to its low signal to noise ratio and higher degree of alignment and orthogonality between its sensors compared to that in the Smart Pebble. In order to collect the data from this system, microDaq wireless Bluetooth (BT-26) data acquisition has been used for that purpose. This acquisition system has 16-channel analog inputs with capability of a maximum sampling rate of 3 kHz. Figure 9 shows the experimental setup and the testing results obtained.

From Figures 8 and 9 it is shown that the new gravity estimation method is capable of filtering out the changes in gravity from the output of the sensors. However, due to accumulation of errors, mainly from the gyroscopes, the

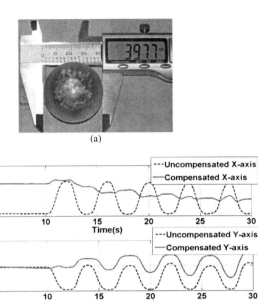

Fig. 8. Testing the new gravity method on the Smart Pebble. (a) Current hardware prototype. (b) Experimental results.

efficiency of the filtering algorithm deteriorates over time. By comparing Figures 8-b and 9-c it is observed that the accumulation of errors in the MAG3D system is slightly better than those in the Smart Pebble. This should be expected since the properties of the MAG3D system in terms of noise and sensors alignments are better than the ones in the Smart Pebble.

With regard to testing the algorithm on the GFINS, the hardware prototype of the cube has been used in this application [18]. This prototype consists of six accelerometers arranged in a cube of length 3cm. After the initial sensors calibration and determining the orientation errors in the sensors as described in [18], changes in gravity around the x-axis has been applied. The data from the sensors have been collected using the same data acquisition system (BT-26). Figure 10 shows the GFINS cube and the testing results. From the results it is seen that the nature of errors in this system is different from the errors found in the GBINS. It is observed that the accumulation of angular errors in this system does not result in deteriorating the efficiency of the gravity estimation method, Instead,

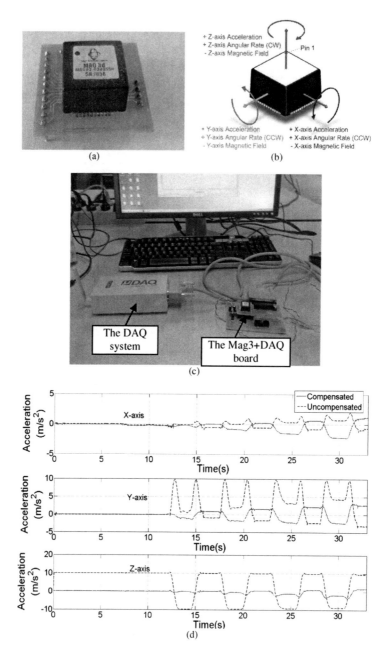

Fig. 9. Testing the new gravity method on the MAG3D INS system. (a) The actual hardware. (b) A schematic of the system. (c) The testing setup. (d) The experimental results.

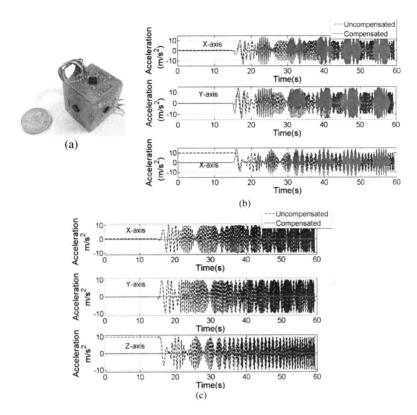

Fig. 10. Testing the new gravity method on the GFINS cube. (a) The actual hardware. (b) The experimental results before applying low pass filter. (c) The results after the filter.

it results in increasing the number of oscillations as the time progresses as shown in Figure 10-b. To eliminate such undesirable oscillations, a simple low pass filter has been applied to filter out these oscillations which helps to improve the efficiency of the gravity estimation algorithm. Figure 10-c shows the result of applying the low-pass filter after the algorithm is used.

Another effect examined using the cube was to verify that if the new algorithm could result in reduction of the GFINS errors according to (22). This was done under stationary conditions by placing the cube with its z-axis against the gravity direction. Two gravity models have been used as input to the GFINS cube: (a) the ideal gravitational values and (b) the gravity values estimated using the new method. Figure 11 shows that the offset errors from the final GFINS outputs which resulted from the new method has a better performance than the offset when using ideal gravity values. With these results, equation (22) is proven to be valid and there is a correlation between reducing the GFINS errors and using the new compensation method to estimate the gravity values.

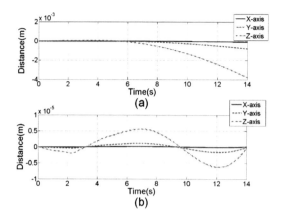

Fig. 11. Offset output of the GFINS cube hardware prototype. (a) Ideal gravity values considered. (b) Gravity is measured using the new model.

8 Conclusions

A new algorithm for estimating and compensating the gravitational values directly from the INS sensors has been described. The algorithm has been applied to Gyro-Based and Gyro-Free INS methods by developing mathematical models for measuring and compensating gravity in each method. The models were validated using simulation and then tested using hardware prototypes.

In Gyro-Based INS, the new gravity compensation models work efficiently for the first few seconds of operation and then the performance deteriorates with time due to accumulation of errors in sensors. Error analysis of the effect of the new gravity estimation algorithm has been studied and it is found that the new method may result in higher error accumulation in GBINS system especially when considering gyroscopes errors. On the other hand, the algorithm improves the overall performance of the GFINS system when considering location and orientation errors of the sensors in the Gyro-Free INS cube.

References

[1] Chen, J.H., Lee, S.C., DeBra, D.B.: Gyroscope Free Strapdown Inertial Measurement Unit by Six Linear Accelerometers. Journal of Guidance, Control and Dynamics 17, 286–290 (1994)
[2] Chin-Woo, T., Park, S., Mostov, K., Varaiya, P.: Design of gyroscope-free navigation systems. In: IEEE Proceedings on Intelligent Transportation Systems, 2001, pp. 286–291 (2001)
[3] Tan, C.W., Park, S.: Design of accelerometer-based inertial navigation systems. IEEE Transactions on Instrumentation and Measurement 54, 2520–2530 (2005)

[4] Ding, M., Zhou, Q., Wang, Q., Wang, C.: Feasibility Analysis of Accelerometer Configuration of Non-gyro Micro Inertial Measurement Unit. In: CES/IEEE 5th International on Power Electronics and Motion Control Conference, IPEMC 2006, vol. 3, pp. 1–4 (2006)
[5] Hansson, G., Asterland, P., Holmer, N.G., Skerfving, S.: Validity and reliability of triaxial accelerometers for inclinometry in posture analysis. Medical and Biological Engineering and Computing 39, 405–413 (2001)
[6] Luinge, H.J., Veltink, P.H.: Measuring orientation of human body segments using miniature gyroscopes and accelerometers. Medical and Biological Engineering and Computing 43, 273–282 (2005)
[7] Rogers, R.M.: Applied mathematics in integrated navigation systems, 3rd edn. American Institute of Aeronautics and Astronautics, Blacksburg (2007)
[8] Titterton, D.H., Weston, J.L.: Strapdown Inertial Navigation Technology, vol. 5. IEEE, London (1997)
[9] Vanderwerf, K.D.: Inertial Navigation with gravity deflection compensation, United States Patent, Patent No:5774832 (1998)
[10] Kwon, J.H., Christopher, J.: Gravity Requirements for Compensation of Ultra-Precise Inertial Navigation. The Journal of Navigation 58, 479–492 (2005)
[11] Jircitano, A., Dosch, D.E.: Gravity Aided Inertial Navigation System, United States Patent, Patent No:5339684 (1994)
[12] Hsu, D.Y.: Method for determining gravity in an inertial navigation system, United States Patent, Patent No:6073077 (2000)
[13] Baerveldt, A.J., Klang, R.: A low-cost and low-weight attitude estimation system for an autonomous helicopter. In: IEEE International Conference on Intelligent Engineering Systems, INES 1997, Budapest, Hungary, pp. 391–395 (1997)
[14] Akeila, E., Salcic, Z., Kularatna, N., Melville, B., Dwivedi, A.: Testing and calibration of smart pebble for river bed sediment transport monitoring. In: IEEE Sensors, 2007, Atlanta, Georgia, USA, pp. 1201–1204 (2007)
[15] Chin-Woo, T., Sungsu, P.: Design of accelerometer-based inertial navigation systems. IEEE Transactions on Instrumentation and Measurement 54, 2520–2530 (2005)
[16] Akeila, E., Salcic, Z., Swain, A.: Direct gravity estimation and compensation in strapdown INS applications. In: 3rd International Conference on Sensing Technology, ICST 2008, pp. 218–223 (2008)
[17] Memsense, Mag 3D, Triaxial Magnetometer, Accelerometer, Gyroscope Analog Inertial Sensor, Mag3D datasheet
[18] Akeila, E., Salcic, Z., Swain, A.: Implementation, Calibration and Testing of GFINS Models Based on Six-Accelerometer Cube. In: IEEE Region 10 Conference, Hyderabad, India (2008)

Model-Based Phasor Control of a Coriolis Mass Flow Meter (CMFM) for the Detection of Drift in Sensitivity and Zero Point

H. Röck and F. Koschmieder

Institute for Automation and Control Engineering
Christian-Albrechts-University of Kiel
Kaiserstr. 2, 24143 Kiel, Germany
hr@tf.uni-kiel.de

Abstract. Coriolis Mass Flow Meters (CMFM) are directly measuring mass flow, by analyzing the induced oscillation of the measuring pipe. The accuracy is high and the sensitivity regarding other features of the flow to be measured like heat conductivity, heat capacity and viscosity is rather small. CMFM available on the market often show drifts in zero point and sensitivity during normal operation. The well known procedure for online determination of zero point and sensitivity [7], [8] is very sensitive regarding the compensation forces to be calculated and requires a cyclic state transition of the meter which is rather time consuming (about 2 to 3 seconds). The new approach presented in this paper using a phasor model of the meter together with an observer for the phasor variables is a major contribution to enhance the performance of the control system, as the cyclic stimuli applied to the meter are now interpreted as disturbances acting on the control loop. The phasor controller is able to compensate the disturbances much faster than in the previous approach.

1 Introduction

Due to the measurement principle, the accuracy of Coriolis Mass Flow Meters (CMFM) is very high. In addition to measurement of mass flow it is also possible to measure density with high accuracy. The main element of the CMFM is a straight pipe (fig. 1) rigidly connected to a supporting pipe. As the electromagnetic sensors measuring the velocity of the pipe are not an obstacle to the flow, the loss in pressure is very low. The oscillation of the pipe is stimulated via two actuators situated symmetrically to the middle of the pipe. When stimulating the pipe with harmonic forces f_a and f_b, identical in phase and amplitude (fig. 2), the pipe will oscillate in its 1st eigenmode, provided there is no mass flow ($\dot{m} = 0$). The induced oscillation of the pipe is measured by two electromagnetic velocity sensors at the locations a and b. As long as there is no mass flow $\dot{m}$ the measured velocity signals v_a and v_b are in phase as well. The interaction of the forced oscillation of the pipe in its 1st eigenmode and the flow $\dot{m} > 0$ will induce Coriolis forces having opposite

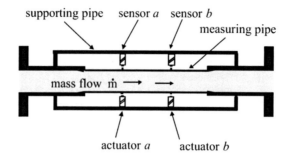

Fig. 1. Principle design of the CMFM

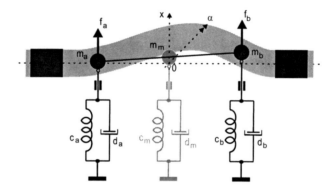

Fig. 2. Equivalent mechanical oscillation system of the CMFM

directions in the up- and downstream part of the pipe. This couple of forces forming a harmonic momentum is stimulating the pipe in its 2nd eigenmode. However, due to asymmetries in the assembly of the CMFM as well as with asymmetries in geometry and material properties, the 2nd eigenmode will always be stimulated even in the case of $\dot{m} = 0$. The superposition of the oscillations in 1st and 2nd eigenmode will result in a phase shift of the velocity signals at the locations a and b. As the phase shift is proportional to the mass flow, it is used as a representation of the mass flow.

2 Statement of the Problem

Industrial CMFM available on the market place nowadays have very good zero stability due to numerous improvements of the mechanical design. Changes in sensitivity however are still occurring and have to be detected and corrected. A model-based analysis using lumped parameters is able to explain these drifts through changes in parameters of the model. For example changes in temperature or in pressure, having impact on Young's modulus, will lead to changes in the coupling of both of the eigenmodes. In this paper a model-based approach and a new control scheme is presented to identify the internal couplings of the CMFM, enabling the detection of drifts in sensitivity and zero point. A model-based analysis of the couplings, explaining the changes in parameters of the lumped parameter model through external conditions (changes in temperature, changes in mounting conditions, strain etc.), will lead to a detailed self-diagnosis of the device in the near future.

3 Lumped Parameter Model

To describe the oscillations of the CMFM in the 1st and 2nd eigenmode, a simple model using lumped parameters is deployed (fig. 3).

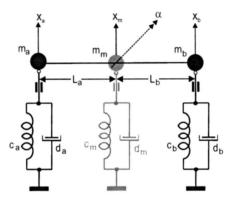

Fig. 3. Lumped parameter model of the CMFM

The oscillation in the 1st eigenmode is represented by a mass damper system with equivalent mass m_m, equivalent spring constant c_m and equivalent damping constant d_m assigned to the middle of the pipe. The equivalent mass m_m thus accounts for a purely translational movement of the pipe. In addition to the translational movement, the oscillation in the 2nd eigenmode is modeled as a revolving movement α of the equivalent masses $m_{a,b}$ with equivalent spring constants $c_{a,b}$ and damping constants $d_{a,b}$ with pivotal point in the center of the pipe.

Superposing the translational movement of the equivalent mass m_m in the center and the rotation α, interpreted as an antiphase movement of the equivalent masses m_a and m_b, results in translational excursions of the pipe at positions a and b described by the new variables η_1 and η_2.

$$\left. \begin{array}{l} x_a \approx x_m + L \cdot \alpha \\ x_b \approx x_m - L \cdot \alpha \end{array} \right\} \Rightarrow \begin{array}{l} x_m \approx \frac{1}{2}(x_a + x_b) := \eta_1 \\ L \cdot \alpha \approx \frac{1}{2}(x_a - x_b) := \eta_2 \end{array} \tag{1}$$

It is straightforward to interpret the variables η_1 and η_2 in the following way: The variable η_1 represents the common mode excursion of the equivalent masses m_m, m_a and m_b, the oscillation in the 1st eigenmode is often called as drive mode. The antiphase excursion of the equivalent masses m_a and m_b with respect to the center of the pipe modeled by the variable η_2 represents the oscillation in the 2nd eigenmode, often named Coriolis mode.

As the forces f_a and f_b stimulate both of the eigenmodes, we are now able to define new forces f_1 and f_2 by extending the previous idea:

$$f_1 = \frac{1}{2}(f_a + f_b)$$
$$f_2 = \frac{1}{2}(f_a - f_b) \tag{2}$$

which stimulate the 1st and 2nd eigenmode independly.

Using the new coordinates η_1 and η_2 and the Euler-Lagrange formalism we can derive two second order differential equations. These differential equations are coupled and the couplings can be interpreted as asymmetries of the mechanical set up:

$$\begin{aligned}\ddot{\eta}_1 + 2d_1\omega_{01}\dot{\eta}_1 + \omega_{01}^2\eta_1 - k_1 k_a \ddot{\eta}_2 - k_1(k_v + k_{ci}\dot{m})\dot{\eta}_2 - k_1 k_s \eta_2 &= k_1 f_1 \\ \ddot{\eta}_2 + 2d_2\omega_{02}\dot{\eta}_2 + \omega_{02}^2\eta_2 - k_2 k_a \ddot{\eta}_2 - k_2(k_v + k_{cn}\dot{m})\dot{\eta}_2 - k_2 k_s \eta_2 &= k_2 f_2\end{aligned} \tag{3}$$

with

$$\begin{aligned}\omega_{01}^2 &= \frac{c_a + c_b + c_m}{m_a + m_b + m_m} \\ \omega_{02}^2 &= \frac{c_a + c_b}{m_a + m_b} \\ d_1 &= \frac{1}{2}\frac{d_a + d_b + d_m}{\sqrt{(c_a + c_b + c_m)(m_a + m_b + m_m)}} \\ d_2 &= \frac{1}{2}\frac{d_a + d_b}{\sqrt{(c_a + c_b)(m_a + m_b)}} \\ k_1 &= \frac{1}{m_a + m_b + m_m} \\ k_2 &= \frac{1}{m_a + m_b}\end{aligned} \tag{4}$$

and the coupling parameters

$$\begin{aligned}k_a &= m_b - m_a \\ k_s &= c_b - c_a \\ k_v &= d_b - d_a\end{aligned} \tag{5}$$

As the coupling k_v from 1st to 2nd eigenmode and 2nd to 1st eigenmode is enhanced by mass flow according to the Coriolis principle, the velocity coupling k_v in eq. (5) is replaced by k_{vij}, now depending on mass flow:

1st to 2nd eigenmode: $\quad k_{v12} := k_v + k_{cn} \cdot \dot{m}$

2nd to 1st eigenmode: $\quad k_{v21} := k_v + k_{ci} \cdot \dot{m}$. \hfill (6)

The constants k_{cn} and k_{ci} are device dependent and mathematically summarize the distributed Coriolis forces acting at the locations a and b.

The state space description of this model with the according constants is given in eqs (7) and (8).

$$\begin{bmatrix} \dot{\eta}_1 \\ \ddot{\eta}_1 \\ \dot{\eta}_2 \\ \ddot{\eta}_2 \end{bmatrix} = \begin{bmatrix} 0 & 1 & 0 & 0 \\ -\omega_{01k}^2 & -2d_1\omega_{01k} & k_{s21} & k_{v21} \\ 0 & 0 & 0 & 1 \\ k_{s12} & k_{v12} & -\omega_{02k}^2 & -2d_{2k}\omega_{02k} \end{bmatrix} \begin{bmatrix} \eta_1 \\ \dot{\eta}_1 \\ \eta_2 \\ \dot{\eta}_2 \end{bmatrix}$$

$$+ \begin{bmatrix} 0 & 0 \\ k_{u11} & k_{u21} \\ 0 & 0 \\ k_{u12} & k_{u22} \end{bmatrix} \begin{bmatrix} u_1 \\ u_2 \end{bmatrix} \tag{7}$$

$$\begin{bmatrix} y_1 \\ y_2 \end{bmatrix} = \begin{bmatrix} 0 & 1 & 0 & 0 \\ 0 & 0 & 0 & 1 \end{bmatrix} \begin{bmatrix} \eta_1 \\ \dot{\eta}_1 \\ \eta_2 \\ \dot{\eta}_2 \end{bmatrix}$$

with

$$\begin{aligned}
u_1 &= f_1 \\
u_2 &= f_2 \\
\omega_{01k}^2 &= k_d(\omega_{01}^2 - k_1 k_2 k_a k_s) \\
\omega_{02k}^2 &= k_d(\omega_{02}^2 - k_1 k_2 k_a k_s) \\
2d_{1k}\omega_{01k} &= k_d[2d_1\omega_{01} - k_1 k_2 k_a(k_v + k_{cn}\dot{m})] \\
2d_{2k}\omega_{02k} &= k_d[2d_2\omega_{02} - k_1 k_2 k_a(k_v + k_{ci}\dot{m})] \\
k_{s12} &= k_d k_2(k_s - k_a \omega_{01}^2) \\
k_{s21} &= k_d k_1(k_s - k_a \omega_{02}^2) \\
k_{v12} &= k_d k_2[(k_v + k_{cn}\dot{m}) - 2k_a d_1 \omega_{01}] \\
k_{v21} &= k_d k_1[(k_v + k_{ci}\dot{m}) - 2k_a d_2 \omega_{02}] \\
k_{u11} &= k_d k_1 \\
k_{u12} &= k_{u21} = k_d k_1 k_2 k_a \\
k_{u22} &= k_d k_2 \\
k_d &= \frac{1}{1 - k_1 k_2 k_a^2}
\end{aligned} \tag{8}$$

The outputs y_1 and y_2 correspond to the velocity signals of the 1st and 2nd eigenmode measured by electrodynamic sensors.

To highlight the various couplings between 1st and 2nd mode, fig. 4 depicts an equivalent block-diagram of the state space description of the CMFM. The transfer functions

$$G_i(s) = \frac{Y_i(s)}{U_i^*(s)} = \frac{k_i s}{s^2 + 2d_i\omega_{0i}s + \omega_{0i}^2} \qquad i = 1, 2 \qquad (9)$$

are describing the oscillation of the measuring pipe in its 1st and 2nd eigenmode. As can be seen from eq. (5), the coupling constants k_a, k_v and k_s are caused by unavoidable asymmetries in the assembly of the CMFM and are associated to couplings proportional to acceleration, velocity and excursion. The distributed Coriolis forces proportional to mass flow and proportional to the velocity $\dot{\eta}_1$ of the driving mode are summarized in the coupling constant k_{cn}, only depending on the geometry of the measuring pipe (see eq. (6)). The impact of $\dot{\eta}_2$ on the drive mode due to mass flow is modeled accordingly by the constant k_{ci}. To summarize the various couplings in the block-diagram between 1st and 2nd mode (compare fig. 4), additional transfer functions $\underline{X}_{12}$ and $\underline{X}_{21}$ are introduced:

$$\underline{X}_{12}(j\omega) = k_{cn}\dot{m} + k_v + j\omega k_a + \frac{k_s}{j\omega} \qquad (10)$$

$$\underline{X}_{21}(j\omega) = k_{ci}\dot{m} + k_v + j\omega k_a + \frac{k_s}{j\omega} \qquad . \qquad (11)$$

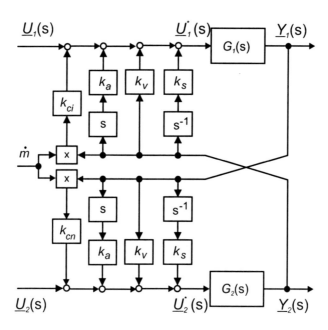

Fig. 4. Block-diagram of the CMFM

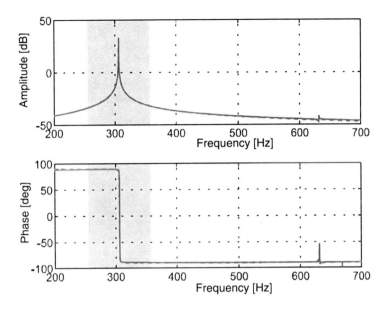

Fig. 5. Measured (solid line) and identified (dashed line) frequency response of $G_{11}(j\omega)$

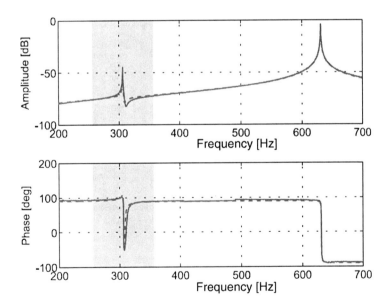

Fig. 6. Measured (solid line) and identified (dashed line) frequency response of $G_{12}(j\omega)$

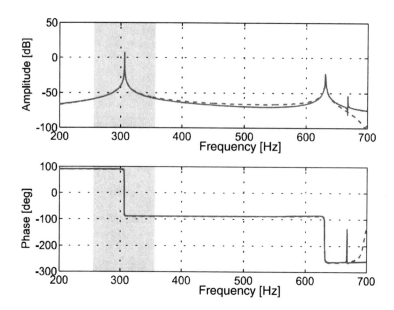

Fig. 7. Measured (solid line) and identified (dashed line) frequency response of $G_{21}(j\omega)$

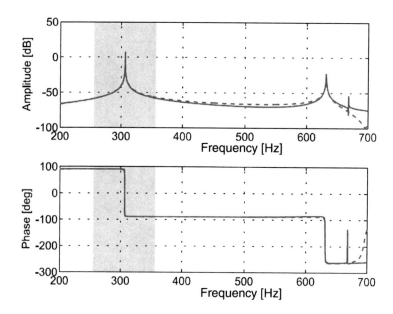

Fig. 8. Measured (solid line) and identified (dashed line) frequency response of $G_{22}(j\omega)$

To validate the mathematical model, the frequency responses G_{ij} of eq. (12) were identified by stimulating the inputs U_1 and U_2 independently with sinusoidal signals in the frequency range of 200–700 Hz, thus containing the eigenfrequencies of the 1st and 2nd mode.

$$\begin{bmatrix} Y_1(j\omega) \\ Y_2(j\omega) \end{bmatrix} = \begin{bmatrix} G_{11}(j\omega) & G_{12}(j\omega) \\ G_{21}(j\omega) & G_{22}(j\omega) \end{bmatrix} \begin{bmatrix} U_1(j\omega) \\ U_2(j\omega) \end{bmatrix} \qquad (12)$$

Identification was done using the Program ELIS from the Matlab Frequency-Domain-Toolbox and is based on a Maximum-Likelihood estimation where the measured frequency response in combination with the estimated variance are used to calculate an unbiased estimate for the coefficients of the four transfer functions. The Bode plots of the identified and measured transfer functions are depicted in figs 5–8 and are in good accordance. The eigenfrequencies of the 1st and 2nd mode occur at about 310 Hz and 630 Hz, respectively. Above the eigenfrequency of the 2nd mode there is a further resonance peak at about 670 Hz that is not explained by the mathematical model. But as the device is operated in a frequency range of about 250–350 Hz only, highlighted in figs 5–8, this modeling error can be neglected.

4 Interpretation of the Sensor Readings in the Complex Plane

Mass flow in a CMFM is proportional to the phase shift of the velocity signals $\underline{V}_a$ and $\underline{V}_b$ at the up- and downstream position of the pipe, where the phasors are represented by underlined capital letters. As the eigenfrequency ω_{01} of the 1st eigenmode is depending on fluid density, the shift in phase $\Delta\varphi = \omega_{01} \cdot \Delta t$ is replaced be the measurement Δt—in the following, this is left out of consideration.

A new interpretation of the sensor readings $\underline{V}_a$ and $\underline{V}_b$ leads to a phasor representation (fig. 9). The sensor readings form a parallelogram with the diagonals representing the sum and the difference of the complex sensor readings, respectively. If a control system accounts for orthogonality of the phasors $\underline{Y}_1$ and $\underline{Y}_2$, the characteristics of the CMFM directly follow from

$$\frac{\omega_{01}\Delta t}{2} = \frac{\Delta\varphi}{2} = \arctan\frac{\text{Im}\{\underline{Y}_2\}}{\text{Re}\{\underline{Y}_1\}} \approx \frac{\text{Im}\{\underline{Y}_2\}}{\text{Re}\{\underline{Y}_1\}} \qquad (13)$$

Thus the control objective is to orthogonalize the velocity signals $\underline{Y}_1$ and $\underline{Y}_2$. Since the CMFM is harmonically stimulated, it is an obvious approach to directly control the phasor signals instead of the sinusoidal time signals. Written in terms of the phasor signals the control objective becomes $\text{Re}\{\underline{Y}_1\}$ is controlled to a constant set point W_{1R}, therefore the mass flow can be represented in two ways:

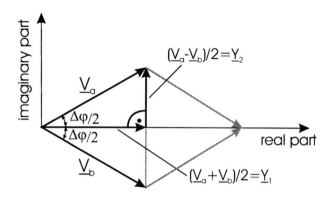

Fig. 9. Phasor signals $\underline{V}_a$ and $\underline{V}_b$ in the complex plane

$$\text{Re}\{\underline{Y}_1\} \equiv W_{1R} \quad (14) \qquad \text{Im}\{\underline{Y}_1\} \equiv 0 \quad (16)$$
$$\text{Re}\{\underline{Y}_2\} \equiv 0 \quad (15)$$

- $\dot{m}$ is proportional to the difference in phase $\Delta\varphi$ of the two output signals $\underline{V}_a$ and $\underline{V}_b$.
- $\dot{m}$ is proportional to the imaginary part of the phasor signal $\underline{Y}_2$ (see eq. (13)).

5 Calculating Sensitivity and Zero Point by Analyzing the Coupling Parameters

If reference values for the phasors as described in the previous sections are given, the phasor control scheme ensures orthogonal phasors $\underline{Y}_1$ and $\underline{Y}_2$ (fig. 10), and in addition, frequency control enables the CMFM to be operated in its 1st eigenfrequency. The different couplings of the 1st and 2nd eigenmode (fig. 4) are denoted as $\underline{X}_{12}$ and $\underline{X}_{21}$, respectively.

From the block-diagram (fig. 10) we derive

$$\text{Re}\{\underline{Y}_1\}\text{Re}\{\underline{X}_{12}\}\text{Im}\{\underline{G}_2\} = \text{Im}\{\underline{Y}_2\} \quad (17)$$

with

$$\text{Im}\{\underline{U}_2\} + \text{Re}\{\underline{Y}_1\}\text{Im}\{\underline{X}_{12}\} = 0 \quad (18)$$

as the imaginary part of the coupling force $\underline{X}_{12} \cdot \underline{Y}_1$ is compensated by control action $\text{Im}\{\underline{U}_2\}$. Taking into account the coupling $\underline{X}_{12}$ from 1st to 2nd mode, the characteristics of the CMFM directly follow from eqs (6) and (13)

$$\frac{\Delta\varphi}{2} \approx \frac{\text{Im}\{\underline{Y}_2\}}{\text{Re}\{\underline{Y}_1\}} = \underbrace{\text{Im}\{\underline{G}_2\}k_{cn}}_{\text{Sensitivity}} \dot{m} + \underbrace{\text{Im}\{\underline{G}_2^*\}k_v}_{\text{Zero point}} , \quad (19)$$

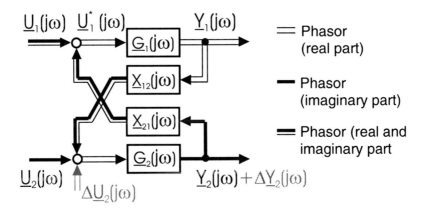

Fig. 10. Block-diagram of the CMFM for $\omega_B = \omega_{01}$

consisting of two expressions, which can be interpreted as sensitivity and zero point. Changes in sensitivity and zero point will always occur simultaneously. Nevertheless zero point stability of commercial CMFM nowadays is very good due to major improvements of the mechanical design reducing the coupling constant k_v ($k_v \approx 0.008$). The sensitivity, however is affected by the viscosity of the fluid and has to be corrected during normal operation of the CMFM. With $\text{Re}\{\underline{U}_1\}$, $\text{Im}\{\underline{U}_1\}$ and $\text{Im}\{\underline{U}_2\}$ as control actions, establishing orthogonality of the phasors $\underline{Y}_1$ and $\underline{Y}_2$, and $\text{Re}\{\underline{U}_2\} \equiv 0$ the mass flow $\dot{m}$ is directly mapped into the output $\text{Im}\{\underline{Y}_2\}$.

Additional information on sensitivity is provided, when a further input $\text{Re}\{\Delta\underline{U}_2\}$ representing a virtual mass flow and interpreted as a deterministic disturbance is periodically induced into the control loop, resulting in

$$\text{Im}\{\underline{G}_2\} = \frac{\text{Im}\{\Delta\underline{Y}_2\}}{\text{Re}\{\Delta\underline{U}_2\}} \quad . \tag{20}$$

As in steady state the control establishes the equality

$$\text{Im}\{\underline{Y}_2\}\text{Re}\{\underline{X}_{21}\} = \text{Im}\{\underline{U}_1\} \tag{21}$$

the velocity coupling k_v can be derived to be

$$k_v = \frac{k_{ci}\text{Im}\{\underline{Y}_2\}^2 + k_{cn}\text{Im}\{\underline{U}_1\}\text{Re}\{\underline{Y}_1\}\text{Im}\{\underline{G}_2\}}{(k_{ci} - k_{cn})\text{Im}\{\underline{G}_2\}\text{Im}\{\underline{Y}_2\}\text{Re}\{\underline{Y}_1\}} \quad . \tag{22}$$

Provided, that the disturbance $\text{Re}\{\Delta\underline{U}_2\}$ is periodically induced into the control loop, new values for the characteristic parameters sensitivity and zero point can be calculated in every cycle. A further option is to implement a compensation method for mass flow measurement. By extending the control scheme using $\text{Re}\{\underline{U}_2\}$ as an additional control action in combination with an additional control objective $\text{Im}\{\underline{Y}_2\} = 0$, the measurement for mass flow is now mapped into Re$\{\underline{U}_2\}$ instead of Im$\{\underline{Y}_2\}$. The implementation of this compensation method is under work.

6 Control Scheme

The advantage of a CMFM phasor model is that all control and reference variables are constant in steady state, whereas in time domain one has to deal with sinusoidal signals. With the assumption that the drive frequency is only slowly varying, the state space model of the CMFM in phasor notation is given in [6]. The control scheme in fig. 11 based on the phasor model has to meet the objectives listed below:

- orthogonality of the phasors $\underline{Y}_1$ and $\underline{Y}_2$ guaranteed by amplitude control
- stimulation of the CMFM in its 1st eigenfrequency ω_{01} ensured by phase control
- fast disturbance rejection (The cyclic stimulation of the 2nd eigenmode is regarded as a deterministic disturbance and the response of the CMFM to this disturbance will provide the necessary information to detect drift in sensitivity and zero point.)

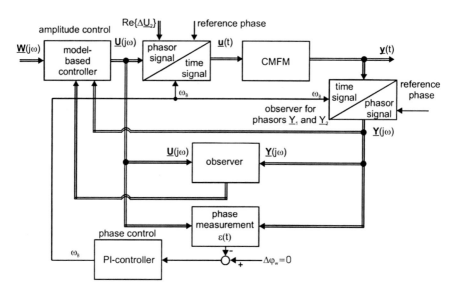

Fig. 11. Model-based phasor control for the 1st and 2nd eigenmode of the CMFM

7 Amplitude Control

The state variable $\underline{\eta}_1$ representing the velocity signal of the 1st eigenmode can be regarded as a phasor in the complex plane, rotating with the angular velocity ω_B (compare fig. 12). The correspondency between phasor description and time domain signal is given by

$$\eta_1(t) = \text{Im}\{(\eta_{1R}(t) + j\,\eta_{1I}(t))e^{j\omega_B t}\} \tag{23}$$
$$\dot{\eta}_1(t) = \text{Im}\{[\dot{\eta}_{1R}(t) - \omega_B\eta_{1I}(t) + j(\omega_B\eta_{1R}(t) + \dot{\eta}_{1I}(t))]\,e^{j\omega_B t}\} \;. \tag{24}$$

If we choose

$$u_1(t) = \text{Im}\{(U_{1R}(t) + j\,U_{1I}(t))\,e^{j\,\omega_B t}\} \;, \tag{25}$$

the MIMO phasor model of the CMFM (four inputs / four outputs) in state space description reads as:

$$\begin{bmatrix}\dot{\eta}_{1R}\\ \dot{\eta}_{1I}\\ \ddot{\eta}_{1R}\\ \ddot{\eta}_{1I}\\ \dot{\eta}_{2R}\\ \dot{\eta}_{2I}\\ \ddot{\eta}_{2R}\\ \ddot{\eta}_{2I}\end{bmatrix} = \begin{bmatrix}0 & \omega_B & 1 & 0 & 0 & 0\\ -\omega_B & 0 & 0 & 1 & 0 & 0\\ -\omega_{01k}^2 & 0 & -2d_{1k}\omega_{01k} & \omega_B & k_{s21} & 0\\ 0 & -\omega_{01k}^2 & -\omega_B & -2d_{1k}\omega_{01k} & 0 & k_{s21}\\ 0 & 0 & 0 & 0 & 0 & \omega_B\\ 0 & 0 & 0 & 0 & -\omega_B & 0\\ k_{s12} & 0 & k_{v12} & 0 & -\omega_{02k}^2 & 0\\ 0 & k_{s12} & 0 & k_{v12} & 0 & -\omega_{02k}^2\end{bmatrix} \cdots$$

$$\cdots \begin{bmatrix}0 & 0\\ 0 & 0\\ k_{v21} & 0\\ 0 & k_{v21}\\ 1 & 0\\ 0 & 1\\ -2d_{2k}\omega_{02k} & \omega_B\\ -\omega_B & -2d_{2k}\omega_{02k}\end{bmatrix} \begin{bmatrix}\eta_{1R}\\ \eta_{1I}\\ \dot{\eta}_{1R}\\ \dot{\eta}_{1I}\\ \eta_{2R}\\ \eta_{2I}\\ \dot{\eta}_{2R}\\ \dot{\eta}_{2I}\end{bmatrix} + \begin{bmatrix}0 & 0 & 0 & 0\\ 0 & 0 & 0 & 0\\ k_{u11} & 0 & k_{u21} & 0\\ 0 & k_{u11} & 0 & k_{u21}\\ 0 & 0 & 0 & 0\\ 0 & 0 & 0 & 0\\ k_{u12} & 0 & k_{u22} & 0\\ 0 & k_{u12} & 0 & k_{u22}\end{bmatrix} \begin{bmatrix}U_{1R}\\ U_{1I}\\ U_{2R}\\ U_{2I}\end{bmatrix}$$

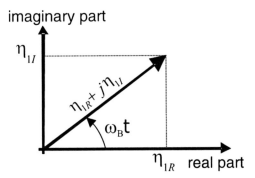

Fig. 12. Phasor representation of $\eta_1(t)$

$$\begin{bmatrix} Y_{1R} \\ Y_{1I} \\ Y_{2R} \\ Y_{2I} \end{bmatrix} = \begin{bmatrix} 0 & 0 & 1 & 0 & 0 & 0 & 0 & 0 \\ 0 & 0 & 0 & 1 & 0 & 0 & 0 & 0 \\ 0 & 0 & 0 & 0 & 0 & 0 & 1 & 0 \\ 0 & 0 & 0 & 0 & 0 & 0 & 0 & 1 \end{bmatrix} \begin{bmatrix} \eta_{1R} \\ \eta_{1I} \\ \dot{\eta}_{1R} \\ \dot{\eta}_{1I} \\ \eta_{2R} \\ \eta_{2I} \\ \dot{\eta}_{2R} \\ \dot{\eta}_{2I} \end{bmatrix} \qquad (26)$$

Orthogonality of the phasors $\underline{Y}_1$ and $\underline{Y}_2$ can be established if the outputs $\text{Re}\{\underline{Y}_1\}$, $\text{Im}\{\underline{Y}_1\}$ and $\text{Re}\{\underline{Y}_2\}$ of the CMFM are kept close to their set points (see eqs (14)–(16)). The necessary control action has to be provided by the control variables $\text{Re}\{\underline{U}_1\}$, $\text{Im}\{\underline{U}_1\}$ and $\text{Im}\{\underline{U}_2\}$. The additional control variable $\text{Re}\{\underline{U}_2\}$ remains unused and is kept to zero.

Analyzing the couplings between 1st and 2nd mode reveals that for this type of operation mass flow is directly mapped into the floating (uncontrolled) output $\text{Im}\{\underline{Y}_2\}$. The controller in combination with an observer is designed as a LQ controller with integral action to avoid steady state errors.

As a virtual mass flow simulated through setting $\text{Re}\{\Delta \underline{U}_2\} \neq 0$ will also be mapped into the corresponding output $\text{Im}\{\underline{Y}_2\}$, a cyclic stimulation of $\text{Re}\{\Delta \underline{U}_2\}$ can be used to gather diagnostic information in order to detect drift in zero point and sensitivity. The control scheme is depicted in fig. 13.

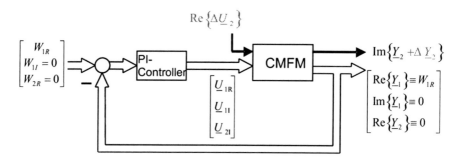

Fig. 13. Control loop / disturbance rejection

8 Phase Control

In single phase flow the damping of the measuring pipe is very small ($d_1 \approx 10^{-5}$), therefore phase control is necessary to ensure that the CMFM is operated in its 1st eigenfrequency, thus maximising the SNR ratio. When stimulating

$$G_1(s) = \frac{k_1 s}{s^2 + 2d_1\omega_{01}s + \omega_{01}^2} = \frac{\underline{Y}_1(s)}{\underline{U}_1^*(s)} \qquad (27)$$

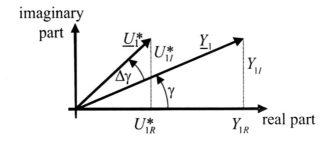

Fig. 14. Phase shift $\Delta\varphi$ between the phasor signals $\underline{Y}_1$ and $\underline{U}_1^*$

in its 1st eigenfrequency ω_{01}, input $\underline{U}_1^*$ and output $\underline{Y}_1$ (compare fig. 14) are in phase, resulting in

$$\tan\gamma = \frac{Y_{1I}}{Y_{1R}} = \frac{U_{1I}^*}{U_{1R}^*}, \quad \Delta\gamma = 0 \qquad (28)$$

or

$$U_{1I}^* Y_{1R} - U_{1R}^* Y_{1I} = 0 \quad . \qquad (29)$$

During transients or when the CMFM is stimulated above or below its 1st eigenfrequency ω_{01}, this equation will not hold. As phase is only defined in steady state, an auxiliary measure

$$\varepsilon(t) = U_{1I}^* Y_{1R} - U_{1R}^* Y_{1I} \qquad (30)$$

for phase is defined that represents the difference in phase between the input and output of $G_1(s)$ and is equal to zero in steady state if $\underline{U}_1^*$ and $\underline{Y}_1$ are in phase.

A simple PI-controller

$$\omega_B = K_p \left[\varepsilon + \frac{1}{T_i} \int_0^t \varepsilon(\tau) d\tau \right] \qquad (31)$$

is used to control the driving frequency until the error in phase approaches zero.

9 Simulation Results

In fig. 15 a simulation result of the closed loop is presented with the uncontrolled output $Y_{2I} = \text{Im}\{\underline{Y}_2\}$ representing mass flow $\dot{m}$. It can be seen, that an additional stimulation of the CMFM by a phasor signal $\text{Re}\{\Delta\underline{U}_2\}$, interpreted as a disturbance, produces an immediate response $\text{Im}\{\Delta\underline{Y}_2\}$. In a further experiment a change in density is simulated as a step change $\Delta\omega_{01}$ in the eigenfrequency of the 1st mode and results in only minor transients of the controlled variables Y_{1R}, Y_{1I} and Y_{2R} during control action until steady state is reached again.

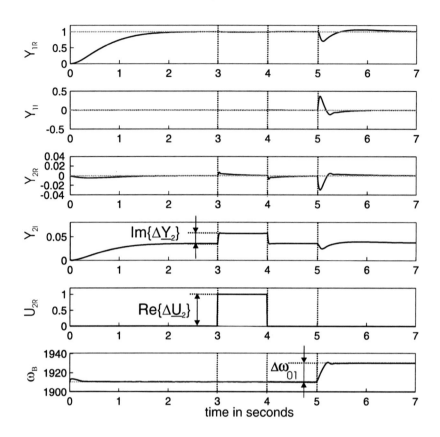

Fig. 15. Phasor control for the 1st and 2nd eigenmode

10 Practical Results

In fig. 16 the cyclic stimulation of the CMFM during normal operation, treated as a disturbance rejection problem, is demonstrated. As can be seen, there is no noticeable impact on the controlled variables $\text{Re}\{\underline{Y}_1\}$, $\text{Im}\{\underline{Y}_1\}$ and $\text{Re}\{\underline{Y}_2\}$. The response of the device to this virtual change in mass flow can be seen in the floating output $\text{Im}\{\underline{Y}_2\}$. According to eq. (19) the sensitivity is given by

$$\text{Im}\{\underline{G}_2\} k_{cn} = \frac{\text{Im}\{\Delta \underline{Y}_2\}}{\text{Re}\{\Delta \underline{U}_2\}} k_{cn} \quad . \tag{32}$$

Thus measuring $\text{Im}\{\Delta \underline{Y}_2\}$ allows to calculate the sensitivity every 1 to 2 seconds without disturbing the mass flow measurement. This is extremely helpful in situations when changes in viscosity will appear that have high impact on sensitivity.

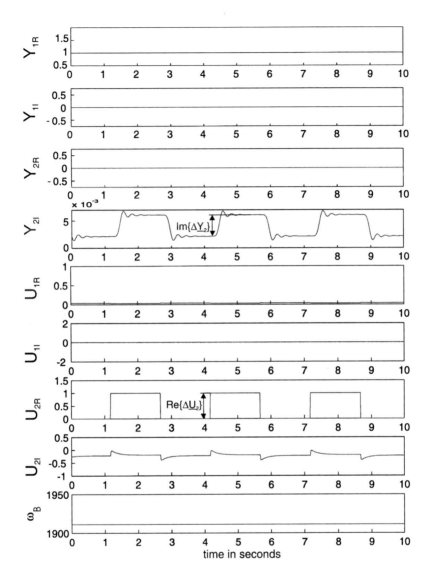

Fig. 16. Response of the closed loop during the cyclic injected disturbances $\text{Re}\{\Delta \underline{U}_2\}$

The control scheme also performs well in situations with multiphase flow. CMFM currently available on the market suffer from poor accuracy or will even stall when multiphase flow enters the measuring device. The accuracy of the measurement in situations with multiphase flow has not yet been investigated, but the proper operation of the device under phasor control has been tested intensively. The results of the experiments carried out are shown

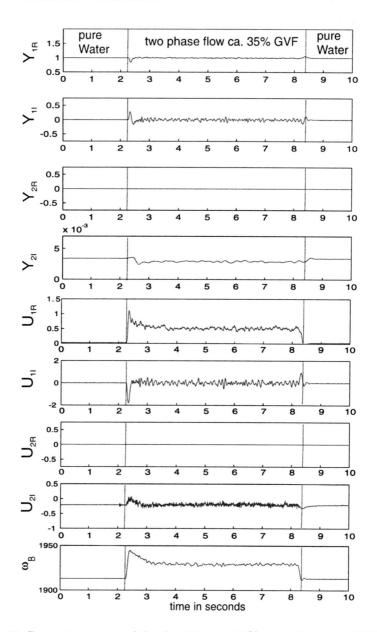

Fig. 17. Dynamic response of the closed loop at 35% gas void fraction (GVF)

in fig. 17. With a gas void fraction of about 35% the damping increases dramatically while the eigenfrequency ω_{01} of the drive mode increases. In order to keep the amplitude at the predefined set point W_{1R}, the control action $\text{Re}\{\underline{U}_1\}$ also increases dramatically while the measurement reading is mapped to the floating output variable Im$\{\underline{Y}_2\}$. Due to a decrease in fluid density, accompanied by an increase in eigenfrequenc ω_{01}, the reading for mass flow Im$\{\underline{Y}_2\}$ decreases as expected.

11 Summary

In this paper a phasor control scheme is presented that allows for online estimation of sensitivity and zero. The state space model of the CMFM in phasor representation has four inputs and four outputs from which three inputs and three outputs are used for MIMO control. The floating (uncontrolled) output represents mass flow as does the phase shift between the sensor readings at location a and b of the measuring device. The input Re$\{\underline{U}_2\}$ is not used for control but can be interpreted as a virtual mass flow measured through Im$\{\underline{Y}_2\}$. Cyclic stimulation of Re$\{\underline{U}_2\}$ thus produces an increase in Im$\{\underline{Y}_2\}$ as if an additional mass flow would pass the measuring pipe and thus allows for online estimation of sensitivity and zero point. As this cyclic stimulation of the CMFM can be interpreted as a deterministic disturbance, the underlying MIMO control scheme has been optimized for disturbance rejection and thus allows for short cycle times.

By extending the control scheme to a four input / four output system, while keeping the floating output Im$\{\underline{Y}_2\}$, representing mass flow, at zero i.e. compensating the Coriolis forces, the mass flow is now mapped into the control action Re$\{\underline{U}_2\}$ and thus realizes a compensation method. The presented control scheme together with the mathematical model of the CMFM is also useful to obtain diagnostic information for example to detect deposits in the measuring pipe at an early stage in order to reduce maintenance costs.

A disadvantage of the presented control scheme lies in the quasi stationary approximation ($\omega_B \approx$ constant) which makes the presented state space model only applicable for slowly varying frequencies ω_B thus limiting the control performance. These limitations however can be overcome when taking into account time varying frequencies ω_B. When doing so, the CMFM model and the feedback control becomes time varying. This concept has already been investigated in simulation studies and reveals that if sufficient control action can be provided, the transients for amplitude and phase control will reach steady state much faster, i.e. within one or two periods of oscillation. This concept will be published in the near future.

References

1. Henry, M.P., Tombs, M.S., Yeung, H.: Coriolis Meter in Two Phase Flows. In: 12th International Conference on Multiphase Production Technology, Barcelona, Spain, May 20-27 (2005)
2. Henry, M.P., Zamora, M.E., Clark, C., Cheesewright, R., Mattar, W.M.: The Dynamic Response of Coriolis Mass Flowmeters. In: Theory and applications, ISA, Houston (October 2004)

3. Kolahi, K., Röck, H.: Density and Pressure Measurement with Coriolis Mass Flowmeters. In: International Conference on Instrumentation, Communication and Information Technology (ICICI), Indonesia, August 3-5 (2005)
4. Röck, H., Koschmieder, F.: A New Method to Detect Zero Drift and Sensitivity of a Coriolis Mass Flow Meter (CMFM) by Using Phasor Control. In: 3rd International Conference on Sensing Technology, Tainan, Taiwan, pp. 518–522 (2008), IEEE Explore No. 10415805, ISBN: 978-1-4244-2176-3, 30.11.-3.12
5. Röck, H., Schröder, T., Kolahi, K., Koschmieder, F.: Flatness-Based Control of a Coriolis Mass Flowmeter. In: ISMTII 2007, Sendai, Japan, September 24-27 (2007)
6. Schröder, T.: Modellgestützte Online-Selbstdiagnose bei einem Coriolis-Massendurchflussmesser mit einem einzigen geraden Messrohr (Model-Based Online-Self-Diagnosis of a Coriolis Mass Flowmeter with a Single Straight Pipe). Dissertation, Institute for Automation and Control Engineering, Christian-Albrechts-University of Kiel (2006)
7. Storm, R., Kolahi, K., Röck, H.: Detection of Zero Shift in Coriolis Mass Flowmeters. In: Proceedings of the XVI IMEKO World Congress, Austria, September 25-28, vol. VI, pp. 61–66 (2000)
8. Storm, R., Kolahi, K., Röck, H.: Model-Based Correction of Coriolis Mass Flowmeters. In: 18th IEEE Instrumentation and Measurement Technology Conference, Budapest, Hungary, May 21-23, vol. 2, pp. 1231–1236 (2001)

Wireless Interrogation of a Micropump and Analysis of Corrugated Micro-diaphragms

Don W. Dissanayake[1], Said F. Al-Sarawi[2], and Derek Abbott[3]

[1] Centre for High Performance Integrated Technologies and Systems (CHiPTec),
School of Electrical and Electronic Engineering,
University of Adelaide, Australia
don@eleceng.adelaide.edu.au

[2] Centre for High Performance Integrated Technologies and Systems (CHiPTec),
School of Electrical and Electronic Engineering,
University of Adelaide, Australia
alsarawi@eleceng.adelaide.edu.au

[3] Centre for Biomedical Engineering (CBME),
School of Electrical and Electronic Engineering,
University of Adelaide, Australia
dabbott@eleceng.adelaide.edu.au

Abstract. In this chapter, Surface Acoustic Wave (SAW) device based wirelessly operated, batteryless and low-powered microdiaphragm structure is investigated. These diaphragms are intended to establish the actuation mechanism for micropumps and similar flow control devices. The actuation method of the diaphragm relies on the electrostatic coupling between the diaphragm and the output Inter Digital Transducer (IDT) of the SAW device. The theory governing the SAW device based novel actuation mechanism is elaborated. A Finite Element Model (FEM) is developed and analysed using ANSYS tools. Different design methods are considered to enhance the deflection of the diaphragm for a low control voltage. As such, inclusion of different types of corrugations, and selection of different bio-compatible materials for various sections of the diaphragm are analysed. Deflection of the diaphragm is obtained as a function of the electric potential at the output IDT of the SAW device and compared with results obtained from published research. Corrugation types such as pure sinusoidal, arc sinusoidal and toroidal types are included in the analysis. Effective meshing requirements that are specific to the presented model are considered and a mesh is developed to achieve converged results. Results show that the use of corrugations around a square-shaped diaphragm with carefully chosen materials results in better performance than that of a flat diaphragm.

Keywords: MEMS, SAW device, Wireless, Electrostatic, Diaphragm, ANSYS, FEA.

1 Introduction

Micropumps represent a significant challenge in the design of low-powered actuation mechanisms. Generally, valve-less micropumps are considered very attractive as they are low cost devices due to their simple structure [1]. Moreover, micropumps that are targeted for use as *in-vivo* applications, wireless and batteryless operation is highly desired. In these low-powered micropumps, the performance of the diaphragm is highly important.

One approach to address the low–stress high–displacement requirement is the utilization of corrugation technique, and is considered to be an effective way to alleviate the residual stress in diaphragm [2]. Corrugated diaphragms have been verified to be capable of releasing the built-in stress, thereby increasing the mechanical sensitivity of the diaphragm and reducing the irreproducibility [3]. This results in requiring a low actuation voltage to drive microdiaphrgm based structures. Various types of corrugation profiles such as sinusoidal, toroidal, sawtooth, and trapezoidal are being used for MEMS applications [2, 4, 5], and are also discussed in this research.

Apart from micro-dimensions and material properties, bio-compatibility is of great importance in the design of such a diaphragm in order to achieve the optimal performance and expected long term operation. This novel micropump is mainly intended for *in-vivo* applications with wireless interrogation such as for drug delivery and micro-dosing, where small quantities of fluid are targeted to a specific location in the body.

In some cases, even with special surface treatments, MEMS devices can still be treated as external intruders and attacked by the immune system [1]. Hence over time, a corrupted medical device can malfunction. Sometimes an avascular fibrous capsule (made up of fibre containing tissue that lacks blood supply) can build up around the implanted device [6]. In addition, a certain degree of toxicity to tissues is strictly forbidden and this has to be considered in the initial phase in order to assign specific materials [1]. Moreover, the mechanical stress induced on tissue by the micropump's dynamics needs to be reduced as well.

In this chapter, theoretical and Finite Element Analysis (FEA) of a corrugated microdiaphragm is presented. Section 2 explains the wirless interrogation and the operation of the proposed micropump structure. Then, Section 3 provides a detailed theoretical analysis of the actuation mechanism for the diaphragm and also highlights the importance of FEM in complex microfluidic structure analysis. Section 4 in this chapter discuss about the FEM and effective mesh generation techniques specific to the presented corrugated diaphragm model. The results achieved through FEA are presented and discussed in Section 5.

2 Wireless and Low–Power Actuation

2.1 SAW Devices

As shown in Fig. 1, a Surface Acoustic Wave (SAW) device consisting of a solid substrate with input and output IDTs [7]. SAW devices are widely used in MEMS applications that require secure, wireless, and passive interrogation [7]. These devices are based on propagation of acoustic waves in elastic solids and the coupling of these waves to electric charge signals via input and

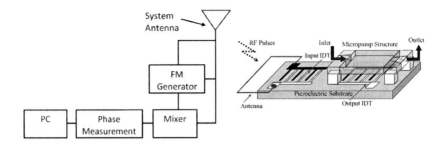

Fig. 1. Low–power, wireless interrogation unit for the SAW device based micropump. SAW device consist of a piezoelectric substrate, input IDT, and output IDT. Input IDT is connected to a micro-antenna for wireless communication. The micropump structure is placed on top of the output IDT of a coded SAW device (SAW correlator), which securely controls the microdiaphragm.

output Inter Digital Transducers (IDT) deposited on a piezoelectric material. Since these are mostly used for wireless applications, a micro–antenna is need to be attached to the input IDT.

Requirement of active electronics to incorporate the security features, hence need of high operating power, and the inclusion of a battery with the implantable device, are considered to be some of the limitations exist in currently available implantable devices [1, 7, 8]. The battery powered implants inherit the added disadvantages such as additional mass and size of the device, and the need to replace the battery.

However, the ability to store energy in SAW, and the possibility of handling high frequencies enables the construction of passive, low–cost devices for wireless communication with the capability to be interrogated by RF signals [9]. Moreover, by using a SAW correlator instead of a standard SAW device, the security feature can be incorporated to the wireless interrogation, with no active electronics being added to the device. The power required to operate the device is obtained from the interrogating RF signal. Therefore, the device is battery-less and become more feasible for implants due to small size and mass. Considering all these factors, this SAW device based novel actuation mechanism is adopted and developed in this research for microdiaphragm analysis.

The micropump structure is designed by integrating a SAW device and a micropump chamber, which causes the microfluidic modulation. The pumping chamber is a suspended structure above the output IDT of the SAW device as shown in Fig. 1. The chamber consists of two diffusers to allow fluid flow, a thin conductive diaphragm , and the surrounding walls to form required enclosure.

The diaphragm actuation principle is as follows. The input IDT converts the incoming RF signal by the inverse piezoelectric effect into SAWs, which propagate along the planar surface of the piezoelectric substrate. At the output IDT, the SAWs are reconverted to an electrical signal as a result of the piezoelectric effect. Since the conductive diaphragm is suspended a few micrometers above the output IDT (as shown in Fig. 2), an electrostatic field is generated between the output IDT and the diaphragm. The resulting electrostatic force deflects the diaphragm. Theory associated with the electrostatic force generation is exploited in Section 3.

3 Theoretical Analysis of Actuation

In most of the microdiaphragm based micropump mechanisms, the deflection of the diaphragm is very small compared to the typical length of the diaphragm. Therefore bending theory of plates is applicable and the transverse deflection for the microdiaphragm can be expressed as [10, 11]:

$$D\nabla^4 W_D + \rho_D t_D \frac{\partial^2 W_D}{\partial t^2} = F - P, \tag{1}$$

where the bending stiffness $D = \frac{E t_D{}^3}{[12(1-\nu^2)]}$, E is the modulus of elasticity, t_D is the diaphragm thickness and ν is the Poisson ratio of the diaphragm material. Moreover, W_D is the deflection in the pump diaphragm, ρ_D is the density of the diaphragm material and ∇^4 is the two dimensional double Laplacian operator. Here, F is the actuating electrostatic force acting on the diaphragm as shown later in Equation 2, while P is the dynamic pressure imposed on the diaphragm by the fluid as shown in Equation 8. The existence of these properties in this context is considered in detail in the following sections.

3.1 Electrostatic Force Generation

In electrostatic actuation, the electrostatic force applied on electrostatic plates can be described using the parallel plate capacitor effect [1] as

$$F = \frac{1}{2} \frac{\varepsilon A \Phi^2}{(h - W_D)^2}, \tag{2}$$

where ε is the dielectric coefficient of the medium between the plates, A is the effective plate area, W_D is the instantaneous deflection of the microdiaphragm in x_3 direction, h is the initial plate spacing, and Φ is the applied electric potential between the plates.

In order to carry out the analysis to derive an expression for the resultant electrostatic force, following assumptions are made of the model and the analysis, as well as simplifications to both being mentioned. For the dimensions chosen for this proposed device, $c/l \approx 100$ GHz, where l is the largest characteristic dimension of the structure (~ 3 mm) and c is the speed of light. As the chosen signal frequency of SAW is in the range of 50 – 100 MHz

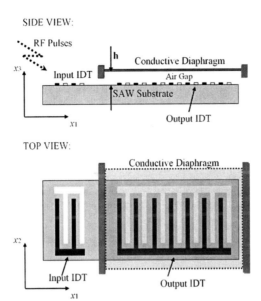

Fig. 2. Side view: The diaphragm is placed above the output IDT of the SAW device, which is separated by an air-gap (h). Top view: The area of the diaphragm is larger than the effective area of the output IDT, hence allows more deflection as the stress levels at the central area of the diaphragm is less.

($\ll c/l$), generated electromagnetic coupling effects can be safely discarded and so the electric field can be treated as quasi-static [12, 13].

Additionally in reality, the electric field lines produced by the positive IDT fingers terminate either at the negative IDT fingers or at the grounded diaphragm. For simplicity however, the effect of the electrostatic coupling between the IDT fingers, as well as the fringe capacitances (between the electrodes and the diaphragm), is discarded in this analysis.

As was previously reported, the electrostatic force is generated due to the time varying electric potential at the output IDT and the conductive diaphragm [14]. Moreover, subjected to the above mentioned assumptions and simplifications, the authors have shown that the electric potential at the output IDT region is a combination of both the electric potential at the IDT and the electric potential at the IDT finger gaps [15, 16], which can be expressed as

$$\Phi(x_1, x_3, t) = \begin{cases} \Psi, & \text{for} \quad n \leq x_1 \leq (\frac{1}{4}+n)\lambda \\ \Omega, & \text{for} \quad (\frac{1}{4}+n)\lambda < x_1 < (\frac{1}{2}+n)\lambda \\ -\Psi, & \text{for} \quad (\frac{1}{2}+n)\lambda \leq x_1 \leq (\frac{3}{4}+n)\lambda \\ -\Omega, & \text{for} \quad (\frac{3}{4}+n)\lambda < x_1 < (1+n)\lambda \end{cases}, \quad (3)$$

where $\Psi = \dfrac{2T}{\pi^2} \Phi(\dfrac{\lambda}{8}, x_3, t)$, $\Omega = \Phi(x_1, x_3, t)$ and $n = 0, 1, 2, ..., N_p - 1$

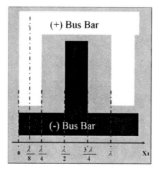

Fig. 3. IDT finger representation for one wavelength (λ) with a metalisation ratio of 0.5

$$\text{for } \Phi(x_1, x_3, t) = \left[\sum_m C_m \alpha_4^m e^{ikb^m x_3} \right] . e^{ik(x_1 - vt)}.$$

Here N_p is the total number of finger pairs in the output IDT, and λ is the SAW wavelength. C_m values are the weighting coefficients of these electric potential equations and are defined based on the mechanical and electrical boundary conditions of the system [15]. These α_4^m values are linear coefficients that are dependent on the decaying constant b, and $m = 1, 2, 3, 4$. v is the SAW velocity in the piezoelectric substrate in the x_1 direction, k is the wave number, and $T \, (= \lambda / v)$ is the time period of the SAW. The coordinate system used in these equations is defined in Fig. 2 and the IDT geometry for one wavelength (one period) of the SAW is shown in Fig. 3.

Due to the periodic nature of the propagating waves and the placement of the IDTs, the electrostatic force analysis is initially carried out only for a single period, and then extended to the whole structure. The single period placement of the output IDT is shown in Fig. 3.

Within one wavelength of the IDT, the analysis is carried out in two parts; one part considering the space above the output IDT electrodes ($0 \le x_1 \le \frac{\lambda}{4} \cup \frac{\lambda}{2} \le x_1 \le \frac{3\lambda}{4}$) and other, the space above the output IDT gap ($\frac{\lambda}{4} < x_1 < \frac{\lambda}{2} \cup \frac{3\lambda}{4} < x_1 < \lambda$). This is due to the IDT fingers consisting of an equipotential distribution for a given time, and the gaps between the fingers consisting of a space varying electric potential distribution in x_1 direction as shown in Equation 3. In this analysis, an IDT with finger width of f_w and finger length of f_l is considered. For a metallisation ration of 0.5 as in Fig. 3, the finger spacing is also f_w.

Section ($0 \le x_1 \le \frac{\lambda}{4} \cup \frac{\lambda}{2} \le x_1 \le \frac{3\lambda}{4}$):
The electrostatic force generated by the electrode finger, which is connected to the positive bus bar $\mathbf{F}_{(+)}$ can be evaluated considering Equations 2 and 3. As a result of the quadratic dependency of force to the applied electric potential, the force generated by the electrode finger, which is connected to the negative bus bar $\mathbf{F}_{(-)}$ is equal to $\mathbf{F}_{(+)}$. Therefore,

$$\mathbf{F}_{(+)} = \mathbf{F}_{(-)} = \frac{\varepsilon_0 f_l f_w}{2(h-W_D)^2} \left(\frac{2T}{\pi^2}\right)^2 \Phi^2(\frac{\lambda}{8}, x_3, t), \qquad (4)$$

where ε_0 is the dielectric coefficient of air.

Section ($\frac{\lambda}{4} < x_1 < \frac{\lambda}{2} \cup \frac{3\lambda}{4} < x_1 < \lambda$):
A slightly different approach is needed to evaluate the electrostatic force generated by the finger gaps. This is because of the space varying electric potential distribution mentioned above. Each finger gap is divided into N_s subdivisions in x_1 direction, so that each subdivision has a width of $\frac{f_w}{N_s}$ and a length of f_l ($\approx$ aperture of the IDT). Combining Equations 2 and relevant range in Equation 3, and after some algebraic simplifications, the electrostatic force generated by each gap can be evaluated as

$$\mathbf{F}_{(\mathbf{gap})} = \frac{\varepsilon_0 f_l f_w}{2N_s (h-W_D)^2} \cdot \left[\sum_j \Phi^2(\frac{\lambda}{4} + \frac{j\lambda}{4N_s}, x_3, t)\right] \qquad (5)$$

for $j = 1, 2, 3,..., N_s$. Therefore, for a distance of single wavelength (λ), the total electrostatic force generated is

$$\mathbf{F}_{(\lambda)} = 2 \left[\mathbf{F}_{(+)} + \mathbf{F}_{(\mathbf{gap})}\right]. \qquad (6)$$

Furthermore, the above results can be used to extend the analysis to the evaluation of the resultant electrostatic force ($\mathbf{F}_{(\mathbf{Tot})}$) generated by an output IDT with N_p pairs of fingers. From Equations 4 – 6,

$$\mathbf{F}_{(\mathbf{Tot})} = \frac{C}{(h-W_D)^2} \sum_j \left[\left(\frac{2T}{\pi^2}\right)^2 \Phi^2(\frac{\lambda}{8}, x_3, t) + \Phi^2(\frac{\lambda}{4} + \frac{j\lambda}{4N_s}, x_3, t)\right] \qquad (7)$$

for $j = 1, 2, 3,..., N_s$ and $C = \frac{\varepsilon_0 f_l f_w N_p}{N_s}$.

3.2 Microfluidic Pressure Variation

In electrostatically actuated micropumps, the fluid flow is considered to be incompressible. Furthermore, non-slip boundary conditions (non-turbulent flow) are assume to exist at pump walls. The governing equations for viscus incompressible fluid flow can be written using Navier-Stokes Equations and mass continuity equation as

$$\rho_L \frac{dV}{dt} = \rho_L g + \mu \nabla^2 V - \nabla P, \qquad (8)$$

$$\frac{\partial \rho_L}{\partial t} + (\nabla \cdot V)\rho_L = 0, \qquad (9)$$

where V is the fluid velocity vector. μ is the viscosity, ρ_L is the density and P is the dynamic pressure of the fluid.

These governing equations show that the electrostatic field between the output IDT and the diaphragm, the deflection of the diaphragm and the flow of the working fluid are always coupled during the pumping action in an electrostatically actuated valveless micropump. In order to analytically determine the deflection of the diaphragm due to excitation force, we need to solve the Equations 1 and 7 – 9 simultaneously.

However such an analytical approach tend to be quite complicated and would require extensive computational effort [17]. Because of the complexity in analysis, which involves electrostatic, structural and fluid field couplings in a complicated geometrical arrangement, FEA can be considered to be more suitable than an analytical system, to study the behaviour of the SAW device based electrostatic micropump. Hence in this research, the FEA of the micropump modules is presented and discussed with the used of the ANSYS simulation tools [18]. Since the implementation of the full corrugated microdiaphragm structure at an integrated FEA would still lead in to complex modelling, the initial detailed analysis of the microdiaphragm as a separate module is considered and presented.

4 Finite Element Modelling of Diaphragms

A coupled-filed analysis is required to model the electrostatic–solid interaction to pursue the Finite Element Analysis (FEA) of the device. Analysis based on load transfer methods are consdered to be more efficient and flexible because different field analysis could be performed independently. ANSYS Multi-field solver (MFX) is available for a large class of coupled analysis problems, and more suitable for MEMS based coupled-field analysis which involves interaction between multiple physics fields [18, 19]. Therefore, ANSYS–MFX solver is to analyse the microdiaphragm performance in this research.

4.1 Model Preparation for Analysis

Node density of the mesh, appropriate element type, and accurate application of boundary conditions are some of the factors that affect the accuracy of FEA. Therefore, extra care is needed in meshing the corrugated microdiaphragm geometry. SOLID95 element type is used for the structural model including 3D corrugations, since it can tolerate irregular shapes without as much loss of accuracy compared to the other solid elements in ANSYS [18]. Additionally, SOLID95 elements are well suited to model curved boundaries, and has capabilities such as plasticity, creep, stress stiffening, large deflection, and large strain capability hence highly suitable for the design of corrugations [20]. Electrostatic air–gap beneath the diaphragm is meshed using SOLID122 element type. SOLID122 elements have compatible voltage shapes and are applicable to 3D electrostatic and time–harmonic electric field analysis [18].

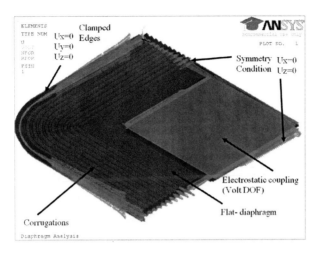

Fig. 4. Boundary conditions applied on the design to reflect the expected constrain. Quarter-symmetry is exploited to reduce the simulation time and CPU usage. The dimensions of the quarter-diaphragm segment are 1000 μm × 5 μm × 1000 μm (Lenght/2 × Height × Width/2). Height of the air-gap is 5 μm.

4.2 Boundary Conditions

The meshed model needs to be constrained by applying appropriate boundary conditions. There are two types of boundary conditions related to this model, namely structural (mechanical constrains) and electric (electrostatic constrains) boundary conditions. These boundary conditions are depicted in Fig. 4.

4.3 Effective Mesh Generation

In electrostatic–solid field coupling, both fields send and receive a physical quantity. In other words, forces get transferred from the electrostatic field to the solid diaphragm and displacements get transferred from the solid to the electrostatic field, which requires a suitable mesh density at the interface.

The effect of the output IDT is incorporated by coupling a set of nodes at the bottom of the air–gap to match the desired IDT pattern and assigning the VOLT Degree of Freedom (DoF) to those nodes. Here, the width of each IDT finger (f_w) is designed to be 10 μm (= $\lambda/4$) for an operating SAW frequency of 50 MHz and SAW velocity of 4000 m/s. Therefore, a smaller element size is needed for an accurate representation of the output IDT as well.

During the initial analysis, simulations were carried out with differently meshed models until the results are converged. This is to highlight the fact that accurate results not only just depend on the node density alone, but also on how well the critical sections are meshed considering the physics associated

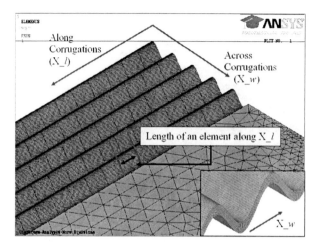

Fig. 5. View of the corrugations in meshed diaphragm model. X_l denotes the direction along the corrugations and X_w denotes the direction across the corrugations towards the mid section.

with the model. Figure 5 explains the different directions considered in each approach to vary the node density in corrugated and flat sections.

First, the node density is changed along X_l direction by varying the length of elements. Then the node density is varied along X_w direction. Once the resolution in corrugated section is established, the element size in the flat section is varied to decide upon an appropriate mesh for flat section, to represent the output IDT accurately. The way these changes affect the accuracy is further discussed in Section 5.1.

4.4 Material Selection for the Device

Taking consideration of the bio-compatibility issues mentioned above in section 1, the material types for different sections of the diaphragm are chosen carefully and their mechanical properties are shown in Table 1. The diaphragm consists of three main sections made from different materials, the

Table 1. Material properties used for the corrugated diaphragm design [21, 22, 23]

Material Property	PI-2610	Si_3N_4	Al
Density (Kg/m^3)	1400	3184	2770
Poisson's Ratio	0.22	0.24	0.33
Elastic Modulus (GPa)	7.5	169	71

flat section, the corrugated section and the conductive thin metal section (can be seen in Figures 4).

Flat Section: Silicon Nitride (Si_3N_4) is chosen for the flat part of the diaphragm, because of its high strength over a wide temperature range, outstanding wear resistance, and good electrical insulation meaning that the fluid in the pumping chamber will be isolated from the electrostatic field.

Corrugated Section: Here, polyimide PI-2610 is chosen for the corrugated section of the diaphragm. The rigid rod polyimide structure of cured PI-2600 products exhibits a desirable combination of film properties such as low stress, low Coefficient of Thermal Expansion (CTE), high elastic modulus compared to other polyimides, and good ductility for microelectronic applications [21]. These attractive features help to generate low-stress high-displacement in the diaphragm and more importantly, facilitate comparatively easy and consistent fabrication of the corrugations around the flat section of the diaphragm.

Conductive Metal Section: Aluminium (Al) is chosen as the material to define metal connections in the micropump device. This is mainly due to the comparatively low elastic modulus and Poisson's ratio, and low cost. However, Gold (Ag) could be used as a better alternative to Al, considering its good conductivity, high bio-compatibility, and the possibility of clean and neat deposition of thin layers ($\ll \mu m$).

In addition to the diaphragm material, the substrate material of the SAW device is crucial to generate the expected SAW mode. Both electrical and mechanical properties of the SAW substrate varies with different crystal cuts in the same material as well as with different piezoelectric materials. Consequently the electro-mechanical coupling between the acoustic wave and the electric potential at the input and output IDTs varies.

5 Simulations and Results Analysis

The dynamic characteristics of the diaphragm depend both on its geometry and material properties. The electrostatic forces will alter the dynamic characteristics of the diaphragm; as it deforms electrostatic forces redistribute, thereby modifying the mechanical loads. Hence structure-electrostatic coupling effects is taken into account during the analysis, using ANSYS–MFX.

5.1 Effect of Mesh Density on Accuracy

Simulation results for the three kinds of mesh variations that was mentioned in Section 4.3 are considered and convergence results are presented in Fig. 6.

In general, results converge as the total number of nodes used for the model increases. However, there is no much variation in results as the node

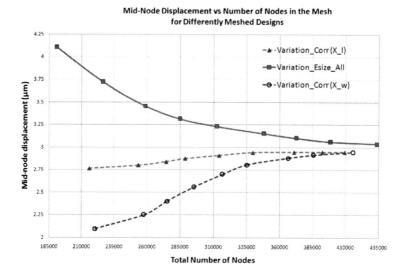

Fig. 6. Mid-node Displacement (MnD) of the diaphragm vs total number of nodes in the mesh. Three scenarios are considered. Variation_Corr (X_l) : MnD as the node density is varied along X_l direction. Variation_Esize : MnD as the element size is varied for flat section. Variation_Corr (X_w) : MnD as the node density is varied along X_w direction. X_l and X_w are defined in Figure 5.

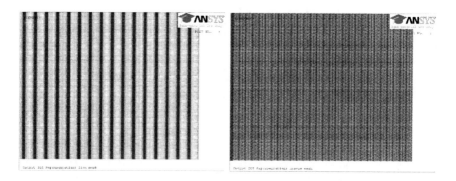

Fig. 7. Left: Output IDT is properly represented in a finely meshed model with an element size of 5 μm. Right: Output IDT is misrepresented in a coarsely meshed model with an element size of 10 μm.

density is changed along X_l. This is because bending stiffness in corrugations along X_l direction is negligible hence does not contribute much towards the sensitivity [24], and addition of extra nodes along that direction does make a little difference.

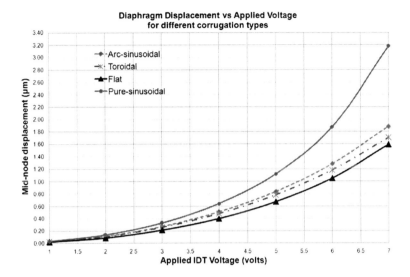

Fig. 8. Comparison of diaphragm displacement of different types of corrugations: Pure Sinusoidal, Toroidal, Approximated sinusoidal with tangent arcs, and Flat diaphragm. Here, all the corrugation parameters [25] are kept constant for consistency.

Whereas, higher variations are observed when the node density is changed along X_w direction. It is known that the tensile rigidity in the radial direction is smaller in a corrugated diaphragm and has a dominant effect towards the overall sensitivity [4]. As a result, a higher node density in X_w direction allows a better representation of the corrugations and produce more accurate results.

Similarly, smaller element sizes in flat section (including the air–gap) has a substantial effect on overall accuracy. As can be seen in Fig. 7, smaller elements allows a good representation of the output IDT structure and also increase the accuracy of the interpolations at the fluid–structure interface during simulations. Therefore as the element size decreases from 15 μm to 3 μm over a few simulations, the convergence is noticed. Once the overall convergence is achieved, the analysis is carried out further to compare diaphragms with different corrugations types.

5.2 Performance Analysis for Different Corrugation Profiles

Fig. 8 depicts mid-node displacements achieved by the diaphragm for different corrugations types. Here the authors have considered pure sinusoidal, arc sinusoidal and toroidal corrugation types for the analysis. As expected, quadratic-shaped curves are observed, in line with the theoretical relationship between the applied electric potential and displacement. More importantly, the corrugated diaphragm in here has high micro-displacements compared to the published research [25, 14]. Based on FEA results it is evident that pure sinusoidal corrugations perform well compared to the other types of corrugations.

In this analysis, von Mises stress is used to predict the yielding of any of the three materials used, under any loading condition. The maximum von Mises stress in this scenario is 1.014 MPa which is much lower that the yield strengths of the selected material. This demonstrates that the diaphragm's deflection is well within the elastic range of the materials used.

6 Conclusion

Analysis of a low-powered microdiaphragm with different corrugation types are presented and the importance of considering bio-compatibility at the design stage is highlighted. A theoretical analysis is carried out explaining the generation of electrostatic force between the SAW device and the diaphragm. By comparing these results with previous results, it is demonstrated that there are several factors that directly contribute in achieving better performance of the diaphragm. The incorporation of corrugated profiles into the diaphragm, a change in the number of corrugation wavelengths, selection of the most suitable diaphragm dimensions and materials for different sections of the diaphragm are some such factors.

As a result of designing the diaphragm to be more flexible, it is been shown to be able to achieve better displacement by carefully choosing materials that have mechanical properties, which are best suited for higher flexibility while maintaining the structural stability. Overall, it is demonstrated that use of polyimide based sinusoidal corrugations on a square shaped diaphragm performs best compared with the other considered types.

Acknowledgements

The authors would like to thank the Australian Research Council (ARC) and the School of Electrical and Electronic Engineering (University of Adelaide) for the funding and the support provided for this research.

References

1. Tsai, N.C., Sue, C.Y.: Review of MEMS-based drug delivery and dosing systems. Sensors and Actuators A 134, 555–564 (2007)
2. Ke, F., Miao, J., Wang, Z.: A wafer-scale encapsulated RF MEMS switch with a stress-reduced corrugated diaphragm. Sensors and Actuators A: Physical 151(2), 237–243 (2009)
3. Wang, W.J., Lin, R.M., Ren, Y., Li, X.X.: Performance of a novel non-planer diaphragm for high-sensitivity structures. Microelectronics Journal 34(9), 791–796 (2003)

4. Giovanni, M.D.: Flat and Corrugated Diaphragm Design Handbook, 2nd edn. Marcel Dekker Inc., New York (1982)
 5. Wang, Y.-G., Shi, J.-L., Wang, X.-Z.: Large amplitude vibration of heated corrugated circular plates with shallow sinusoidal corrugations. Sensors and Actuators A: Physical (in Press), Corrected Proof:1–10 (November 2008)
 6. Butt, O.I., Carruth, R., Kutala, V.K., Kuppusamy, P., Moldovan, N.I.: Stimulation of peri-implant vascularization with bone marrow-derived progenitor cells: monitoring by in vivo EPR oximetry. Tissue Engineering 13(8), 2053–2061 (2007)
 7. Jones, I., Ricciardi, L., Hall, L., Hansen, H., Varadan, V., Bertram, C., Maddocks, S., Enderling, S., Saint, D., Al-Sarawi, S., Abbott, D.: Wireless RF communication in biomedical applications. Smart Materials and Structures 17(015050), 1–10 (2008)
 8. Nguyen, N.T., Huang, X., Chuan, T.K.: MEMS-Micropumps: A review. Transactions of the ASME Journal of Fluids Engineering 124, 384–392 (2002)
 9. Hamsch, M.: An Interrogation Unit for Passive Wireless SAW Sensors Based on Fourier Transform. IEEE Transactions on Ultrasonics, Ferroelectrics, and Frequency Control 51, 1449–1455 (2004)
10. Cui, Q., Liu, C., Zha, X.F.: Simulation and optimization of a piezoelectric micropump for medical applications. The International Journal of Advanced Manufacturing Technology 36(5), 516–524 (2008)
11. Lee, B., Kim, E.S.: Analysis of partly-corrugated rectangular diaphragms using the Rayleigh-Ritz method. Journal of Microelectromechanical Systems 9(3), 399–406 (2000)
12. Horenstein, M.N., Perreault, J.A., Bifano, T.G.: Differential capacitive position sensor for planar MEMS structures with vertical motion. Sensors and Actuators 80, 53–61 (2000)
13. Gantner, A., Hoppe, R.H.W., Köster, D., Siebert, K.G., Wixforth, A.: Numerical simulation of piezoelectrically agitated surface acoustic waves on microfluidic biochips (2005) (visited on, 25/06/2007)
14. Dissanayake, D.W., Al-Sarawi, S., Abbott, D.: Corrugated micro-diaphragm analysis for low-powered and wireless Bio-MEMS. In: 3rd International Conference on Sensing Technology, November 2008, pp. 125–129 (2008); Available in IEEE Explore, ISBN: 978-1-4244-2176-3
15. Dissanayake, D.W., Al-Sarawi, S., Abbott, D.: Surface acoustic wave device based electrostatic actuator for microfluidic applications. In: 2nd International Conference on Sensing Technology, November 2007, pp. 381–386 (2007)
16. Dissanayake, D.W., Al-Sarawi, S.F., Abbott, D.: Electrostatic micro actuator design using surface acoustic wave devices. In: Smart Sensors and Sensing Technology. LNEE, vol. 20, pp. 139–151. Springer, Heidelberg (2008)
17. Bao, M.-H.: Basic mechanics of beams and diaphragm structures. In: Micro mechanical transducers: pressure sensors, accelerometers, and gyroscopes (Hand Book of Sensors and Actators), vol. 8, pp. 23–87. Elsevier, New York (2000)
18. ANSYS Incorporation, http://www.ansys.com/ (visited on, 17/02/2009)
19. Nisar, A., Afzulpurkar, N., Mahaisavariya, B., Tuantranont, A.: Multifield analysis using multiple code coupling of a MEMS based micropump with biocompatible membrane materials for biomedical applications. In: International Conference on BioMedical Engineering and Informatics, March 2008, vol. 1, pp. 531–535 (2008)

20. Lakshmininarayana, H.: Finite Elements Analysis: Procedures in Engineering. Orient Blackswan (2004)
21. HD MicroSystems, Parlin, NJ. PI-2600 LX series, low stress polyimides, Product information and process guidelines (April 2008)
22. Kazaryan, A.A.: Thin-film capacitive pressure transducers which operate during the deformation of products. Measurement Techniques 43(1), 38–43 (2000)
23. Gad-el-Hak, M.: The MEMS Handbook. CRC Press, New York (2002)
24. Füldner, M., Dehé, A., Lerch, R.: Analytical Analysis and Finite Element Simulation of Advanced Membranes for Silicon Microphones. IEEE Sensors Journal 5(5), 857–863 (2005)
25. Dissanayake, D.W., Al-Sarawi, S., Lu, T.-F., Abbott, D.: Design and characterisation of micro-diaphragm for low power drug delivery applications. In: Proc. of SPIE–Active and Passive Smart Structures and Integrated Systems, April 2008, vol. 6928, Article 69282, pp. 1–8 (2008)

Synthesis of Aligned ZnO Nanorods with Different Parameters and Their Effects on Humidity Sensing

Yun Wang[1], John T.W. Yeow[1,*], and Liang-Yih Chen[2]

[1] University of Waterloo, Waterloo, Canada
[2] National Taiwan University of Science and Technology, Taiwan
*corresponding author: jyeow@engmail.uwaterloo.ca

Abstract. In this paper, we report the synthesis of aligned ZnO nanorods with different growing conditions and the effects of the growing parameters on the nanorods humidity sensing characteristics. Six different kinds of samples are prepared with different synthesis parameters, including the zinc source, growing time, with or without PEI and post annealing. All samples show increasing resistance response to increasing relative humidity level at room temperature. This can be explained by the formation of electron depletion layer on the nanorods surface due to the adsorption of water molecules. It is demonstrated that, ZnO nanorods grown from $Zn(Ac)_2$ with shorter growing time and post annealing shows the best overall sensing characteristics. The nanorods without post annealing show much higher sensitivity and faster response, but very fast degradation and unstable response to RH changes.

1 Introduction

As a metal oxide semiconductor, zinc oxide (ZnO) has attracted substantial research interest due to its distinctive optical, electronic and chemical properties. This wide-bandgap material had found its application in various fields, including optoelectronic devices, piezoelectric transducers, transparent conducting film, dye-sensitized solar cell, and solid-state gas sensors, et al [1-4]. D. C. Look [5] and Ü. Özgür [6] have comprehensively reviewed the properties and applications of ZnO as both bulk and nanostructured materials. With the nano-sized ZnO, its morphology and properties can be modified significantly. Therefore, a great deal of attention has been paid to the synthesis and application of nanostructured ZnO, including nanowires, nanorods, nanobelts, nanotetrapod and hollow spheres [7-12].

Metal oxides have always been important gas sensing materials due to its low cost, simplicity and large number of detectable gases [13-14]. Compared with the traditional pores ceramic structure, the nanostructured materials have ultra-high surface-to-volume ratio, hollow structures, small grain size and very large grain boundary areas, which significantly promotes the gas sensing properties. As one of the most important gas sensing metal oxides, nanostructured ZnO has demonstrated its gas sensing capability to a wide range of gases including NO_2, NH_3, H_2, CO, water vapor and organic vapors such as ethanol [15-17].

As one of the most important gas sensors, humidity sensors have widespread applications in many areas —such as in the food processing, semiconductor, and pharmaceutical industries— where humidity is monitored and controlled in real-time to ensure product quality. They are also widely used domestically for monitoring of indoor air quality to ensure human comfort. The focus of this research is to build a humidity sensor with nanostructured ZnO for the application within fuel cell stacks- where the water management problem is crucial to the overall performance [18]. Therefore, humidity sensor that has good chemical and thermal resistance, good sensitivity under high relative humidity level and temperature, and fast response is preferred.

Previously, we reported some preliminary experiment results from synthesized aligned ZnO nanorods and their humidity sensing characteristics [19]. In this paper, more in-depth studies are reported, especially the effect of the synthesis parameters on the nanorods' morphology and humidity sensing behavior. During the preparation, the parameters including synthesis source, time, with or without poly(ethyleneimine) (PEI) and post annealing varied to find the optimal process for humidity sensing.

2 Experimental

2.1 Synthesis of ZnO Nanorods

Aligned ZnO nanorods are synthesized using a hydrothermal method on indium tin oxide (ITO) coated glass substrate. The detailed preparation processes are presented in [19]. For the seed layer preparation, the ITO substrates are irradiated by UV lamp to enhance the wetting of the seed layer. Precipitation deposition in ZnO colloid solution is used to form the seed layer. The ZnO colloid solution is synthesized by method proposed by Hoffmann [20]. The size of the ZnO nanoparticles is approximately 5 nm. The substrates are annealed in air at 550°C for 30 minutes in a furnace to enhance the adhesion and crystallization of the ZnO seed layer on ITO substrates. ZnO nanorods are grown on the seed layer using the hydrothermal method in aqueous solution containing the growing source -zinc acetate ($Zn(AC)_2$) or zinc nitrate ($Zn(NO3)_2$)- and ammonium hydroxide (28 wt% NH_3 in water, 99.99%) at 95°C for certain time. During the growth, PEI can be added to assist growing longer nanorods. After growth, the substrate is removed from the solution, followed by rinsing with DI water, and then dried in air. The length of the ZnO nanorods are controlled by the growth time and long nanorods can be achieved by multi-step growth in the aqueous solution. The as-grown ZnO nanorods can be annealed further in air at 400-550°C for 30-60 minutes to improve the crystallinity of the ZnO nanorods and the interfacial structures. In order to study the effect of post annealing on the humidity sensing property, nanorods with and without post annealing are prepared. Table 1 lists the six different kinds of ZnO nanorods that are prepared varying the growing parameters for this experiment, with the sample 1 as the standard process.

Table 1. ZnO nanorods samples prepared with varying the growing parameters

	Source	Growth Time	Adding PEI	Post annealing
Sample 1	Zn(Ac)$_2$	4	Yes	Yes
Sample 2	Zn(Ac)$_2$	10	No	Yes
Sample 3	Zn(NO$_3$)$_2$	10	No	Yes
Sample 4	Zn(Ac)$_2$	24	Yes	Yes
Sample 5	Zn(Ac)$_2$	24	No	Yes
Sample 6	Zn(Ac)$_2$	24	Yes	No

2.2 Sensor Structure and Measurement Setup

The resistance response of the ZnO nanorods mat array to relative humidity (RH) change is measured in this experiment. Two electrodes contacts are made at each end of the rectangle nanorods area with Al thin film by conducting epoxy. The sensor is then put into a sealed chamber with controllable and stable RH and temperature. The chamber has two gas inlets and one outlet. One inlet connects to extra dry air directly, and the other is connected to a humidification bottle where the extra dry air becomes saturated with water molecules. Two individual mass flow controllers are implemented to control the ratio of the dry and saturated air. As a result, the RH range within the chamber can be controlled continuously from 0% to 100%. A commercial humidity sensor is put into the chamber as a reference to indicate the RH level. Real-time measurement is performed with a LRC/ESR

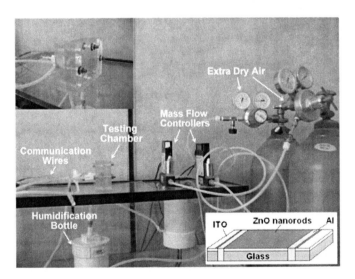

Fig. 1. Experiment setup; insert (up-left) the testing chamber, (lower-right) the humidity sensor's testing structure

meter (PK PRECISION 889A). A LabVIEW program is used to control the mass flow controllers and collect and record the data from both the LCR meter and the reference sensor. Figure 1 shows the measurement setup and the schematic of the sensor's structure in the insert. All the measurements in this paper are done under room temperature.

3 Characterization of the ZnO Nanorods

The crystallinity of the annealed ZnO nanorods is investigated using X-ray diffraction. As shown in Fig. 2, the aligned ZnO nanorods array that is grown on an annealed seed layer substrate shows only the ZnO (000l) peak, indicating that ZnO (0001) planes are oriented parallel to the substrate. These results indicate that individual ZnO nanorods, crystallized along the ZnO [0001] direction, are vertically aligned on the substrate.

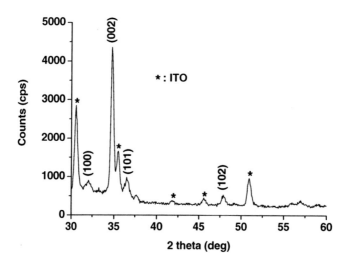

Fig. 2. X-ray diffraction spectroscopy of the ZnO nanorods that are grown on an annealed seed layer substrate

Figure 3 shows the scanning electron microscopy (SEM) images of the ZnO nanorods prepared on ITO/glass with different parameters. A high density of ZnO nanorods grow vertically on the substrate. Depending on the growth condition, the diameters of the nanorods range from 150 nanometer to several hundreds nanometers and the length range from 1 μm to 6.5 μm. Fig. 3(a), (b) shows the image of sample 1, with $Zn(Ac)_2$ as zinc source, PEI in the aqueous solution, 4 hours synthesis time and post annealing at 400°C for 30min. The nanorods show uniform hexagonal structure and an average length of 1 μm. Sample 2 (Fig. 3(c)) and sample 3 (Fig. 3(d)) are grown for 10 hours with different zinc sources and without

Synthesis of Aligned ZnO Nanorods with Different Parameters   261

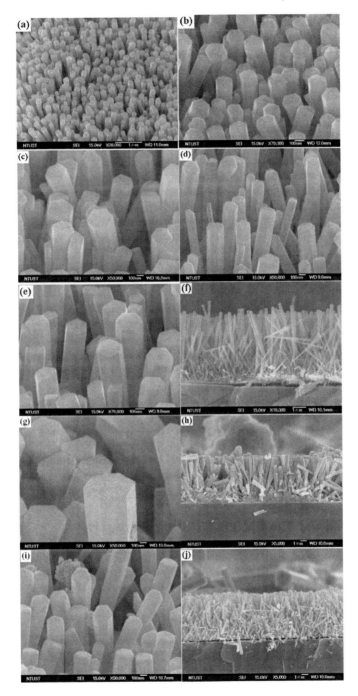

Fig. 3. SEM images of prepared ZnO nanorods, (a) and (b) sample 1; (c) sample 2; (d) sample 3; (e) and (f) sample 4; (g) and (h) sample 5; (i) and (j) sample 6

PEI. Compared with sample 2, which is grown with Zn(Ac)$_2$ as source, sample 3 shows a smaller diameter on average when grown with Zn(NO$_3$)$_2$ as source. PEI is an organic ligand to assist growing longer ZnO rods. It will inhibit the radial growth thus the diameter of the ZnO rods will be smaller. This can be seen when comparing sample 1 (Fig. 3(b)) and sample 2 (Fig. 3(c)). Sample 2 grown without PEI shows larger diameter than sample 1. Sample 4 and sample 5 also shows the same effect. Comparing sample 4 (Fig. 3(e)) and sample 5 (Fig. 3(g)), sample 5 shows a significant increase in the nanorods diameter since no PEI is added during the preparation. Sample 4, 5 and 6 are all grown in Zn(Ac)$_2$ for 24 hours, which results in longer nanorods (6.5 μm for sample 4 as shown in Fig. 3(f)). There is no obvious morphology difference shown in the SEM image between sample 4 and 6. This is due to the post annealing process affecting the surface state of the nanorods instead of their size and shape. Due to this change in surface state, the humidity sensitivity can be improved significantly, which will be discussed in detail in the next session.

4 Humidity Sensing Results and Discussions

To study the influence of the growing condition on the samples' sensing characteristic, their resistance variation between 0% RH and 100% RH at room temperature are measured and compared. The resistance variation is defined as $(\Delta R/R_0)*100\%$, where R_0 is the resistance at 0% RH, and ΔR is the resistance change at given RH to R_0. Figure 4 shows the comparison of the six samples' resistance variations. It can be concluded from the results that:

i. With the increase of humidity level, all samples show increasing resistance.

ii. The longer nanorods show smaller resistance variation (comparing sample 1, 2 and 4). This is probably due to the lower diffusion efficiency of the water molecules into the longer nanorods. Especially when the as-grown aligned nanorods are packed together with a high density, water molecule adsorption may only happen at the top part of the nanorods and thus results in the lower sensitivity.

iii. Comparing sample 2 and 3, it shows that nanorods that are grown with Zn(Ac)$_2$ as zinc source has larger response to humidity change than those with Zn(NO$_3$)$_2$ as zinc source. The difference between the samples is the counter ions in the aqueous solution during preparation: one is CH$_3$COO$^-$ (for Zn(Ac)$_2$) and the other is NO$_3^-$ (for Zn(NO$_3$)$_2$). However, more study is needed to explain the effect of the counter ions on the surface state and humidity sensing property at this point.

iv. Comparing sample 4 and 5, we can see that adding PEI during the preparation does not affect the surface state thus the humidity sensitivity of the nanorods.

v. The post annealing process has significant effect on the humidity sensing characteristic of the nanorods. Nanorods without post annealing show more than 8 times resistance variation than the annealed nanorods (comparing sample 6 and sample 4).

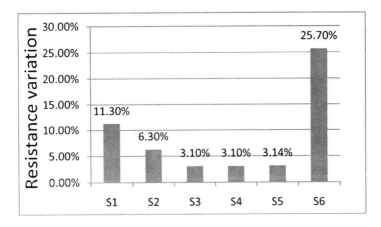

Fig. 4. Comparison of the different samples' resistance variation between 0% and 100% RH (the results are from fresh samples, that is, from the first round test of each sample)

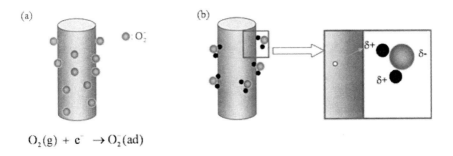

Fig. 5. Schematic of the adsorption of O_2 and H_2O molecules on the surface of ZnO nanorods, (a) the adsorption of O_2, (b) the adsorption of H_2O

For semiconducting metal oxides, it is believed that the gas sensing mechanism under low temperature is dominated by the surface process [21]. For n-type semiconductors, the change of the material's resistance is generally explained by the formation of electron depletion layer on the sensing surface due to the adsorption of oxygen species [22-23]. As oxygen atoms have high electro-negativity, the oxygen atoms will easily trap free electrons from the native n-type ZnO nanorods. When free electrons of ZnO nanorods are trapped, the amount of electrons transported inside the nanorods will decrease, resulting in the increase of their resistance. For water molecules, the mechanism is very similar to oxygen molecules. Since the oxygen atoms will easily attract electrons from the hydrogen atoms in the water molecule, the oxygen will be δ- and hydrogen atom will be δ+. When the H atoms link on the surfaces of ZnO nanorods, they will easily trap electrons from ZnO nanorods and increase their resistance. Similar gas sensing properties of ZnO are observed and reported for other oxidation gas as well. Because the sensing mechanism is a surface process, the sensor's sensitivity highly depends on the surface state of the nanorods. Therefore, the unannealed nanorods, which is not completely oxidized and crystallized on the surface, can provide much more adsorption sites for water molecules, resulting in larger resistance variation under the same condition.

From the results above, we can see that sample 1 and 6 shows better and more interesting results. Therefore, more measurements are performed on these two samples. Fig.6 (a) shows the resistance-RH curve for sample 1. The RH in the

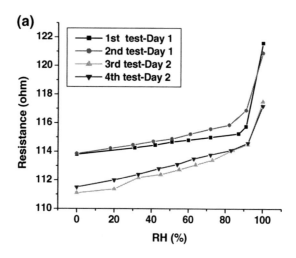

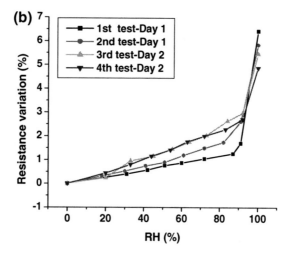

Fig. 6. (a) the Resistance-RH curve of sample 1 for several rounds of test, (b) the corresponding Resistance Variation-RH curve

chamber is changed step-by-step. For each step, the RH level is kept constant until the resistance reading reached a steady level. The measurement is run in two days within a week. It is observed that the baseline resistance reading (R_0) of the sample changes slightly from day to day. However, for the same day, during different round of test, the baseline resistance reading is relatively stable. The sensor shows steady resistance change over a wide range of relative humidity (from 0% to 100%). The sensing curve can be divided into two regions: Region I, 0% RH-90%RH and Region II, 90% RH – 100% RH. For both regions, the sensor shows very linear Resistance-RH relationship. The sensor also shows much higher sensitivity in Region II, which is very preferable for the application in fuel cells (90-100% working environment). Fig.6 (b) shows the corresponding Resistance variation-RH curve of sample 1. Compared to the original resistance variation at 100% RH for the first test (11.3%), the resistance variation at 100% RH slowly decreases from test to test, while the resistance variation at Region I slowly increases and approaching to a steady state. This slowly change is probably caused by the incomplete oxidation of the nanorods surface. Although the sample has been post annealed, there are still unstable adsorption sites on the surface. After several times of test under high humidity level, these unstable sites are slowly oxidized and come to stable state. Therefore, the sensor's response to RH slowly becomes stable. This also explains the better repeatability of the sensor's response on day 2 of the tests.

Fig.7 shows the dynamic resistance response of sample 1 resistance to the changing RH for tests in different days. Comparing these two curves, it can be seen that the dynamic response of the sensor is much more stable on the day 2. After several times of test, the sensor reaches the steady state and shows very good repeatability with fast response. Fig.7 (b) also shows the dynamic resistance response to step change of RH. In Region I (0% RH- 90% RH), the sensor's response time (defined as the time to reach 90% of the total change) is 2 to 3 minutes. In Region II the response time is about 20min, and when RH changes from 0% to 100%, the sensor's response time is 7min with increasing RH. However, the sensor shows very fast response on the recover cycle with a 2min response time.

Compared with sample 1, the unannealed nanorods (sample 6) show very high resistance variation at the first test. However, due to the high density of incomplete oxidation and crystallization sites, the sample shows very fast degradation over time. Three units (#1, #2 and #3) that are made under the same condition as sample 6 are tested for the degradation effect. Figure 8 shows the resistance variation change over time of these three units. #1 and #2 are tested as fresh samples. At the first test, both of them show 25-26% resistance variation at 100% RH. After 10 days, the resistance variation at 100% RH of #1 decreases dramatically to 4.7%. After 20 days from the first test, the resistance variation at 100% RH of #2 decreases to 1%. #3 is tested three month after the first test of #1 and #2. Due to the slowly oxidation in air, the resistance variation of #3 is much lower than #1 and #2. And the change of resistance variation at 100% RH between the first and second test is not as much. The sample's unstable response can also be observed from their dynamic response to changing RH, as shown in Fig.9. The resistance baseline (at 0%) is not as constant as sample 1 from cycle to cycle. However, the

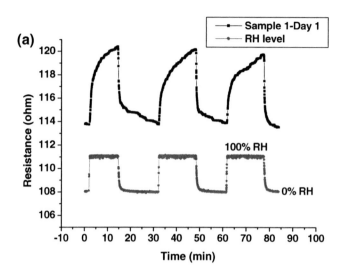

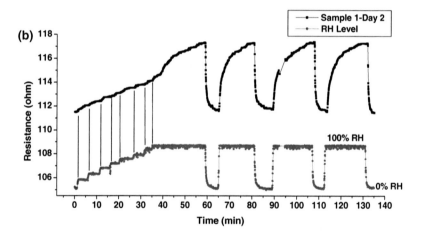

Fig. 7. Dynamic resistance response to change RH of sample 1's for test in different days, (a) day 1, (b) day 2

resistance at 100% RH is relatively stable. And after long time testing, the response slowly turns to be more stable (as shown in the last four cycles in Fig.9). Although sample 6 shows fast degradation over time and unstable responses to RH change, it shows much faster response time compared with sample 1. As seen in Fig.10, the response time for increasing RH from 0% to 100% is 140s and is 60s for the recovery cycle.

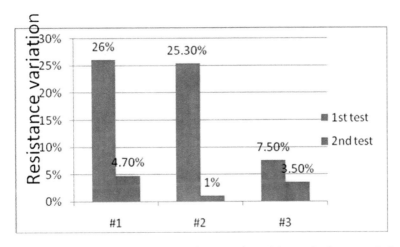

Fig. 8. The resistance variation degradation over time of three units from sample 6

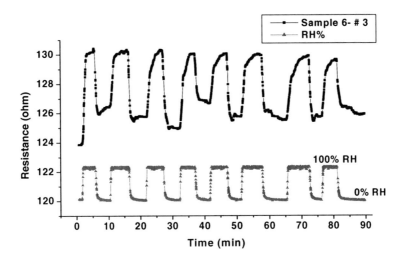

Fig. 9. Dynamic response of Sample 6 (# 3) to RH change cycles between 0% and 100%

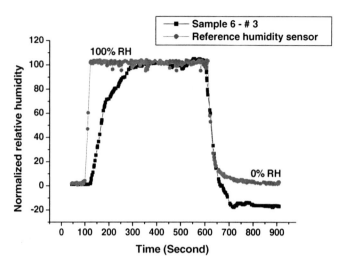

Fig. 10. Comparison of the response time of sample 6 - # 3 with the reference humidity sensor

6 Conclusions

In the paper, we present the synthesis of aligned ZnO nanorods with different conditions and their effect on the nanorods humidity sensing characteristics. Six different kinds of samples are prepared with different growing parameters, including the zinc source, growing time, with or without PEI and post annealing. It is demonstrated that, ZnO nanorods grown from $Zn(Ac)_2$ with shorter growing time and post annealing shows the best overall sensing characteristics. The sensor shows stable and linear resistance response to RH and a high sensitivity under high humidity level, which is very important for the application in fuel cells. Post annealing of the ZnO nanorods shows significant effects on the sensing properties in both positive and negative ways. Compared to nanorods with post annealing, the unannealed nanorods shows both much higher sensitivity and faster response. However, it also shows very fast degradation and unstable response to RH changes. Therefore, more study can be focused on the post annealing temperature and time to optimize the humidity sensing property of the nanorods. Undergoing research also includes humidity sensors with ZnO nanorods laterally aligned between interdigitated electrodes and the humidity sensing property of aligned ZnO nanorods under elevated temperature.

References

[1] Wang, Y.D., Zhou, J., Song, J.H., et al.: Piezoelectric field effect transistor and nanoforce sensor based on a single ZnO nanowire. Nano. Letters 6, 2768–2772 (2006)

[2] Yang, P.D., Yan, H.Q., Mao, S., et al.: Controlled growth of ZnO nanowires and their optical properties. Adv. Funct. Mater. 12, 323–331 (2002)

[3] Law, M., Greene, L.E., Johnson, J.C., et al.: Nanowire dye-sensitized solar cells. Nat. Mater. 4, 445–459 (2005)
[4] Navale, S.C., Gosavi, S.W., Mulla, I.S., et al.: Controlled synthesis of ZnO from nanospheres to micro-rods and its gas sensing studies. Talanta 75, 1315–1319 (2008)
[5] Look, D.C.: Recent advances in ZnO materials and devices. Materials Science and Engineering B 80, 383–387 (2001)
[6] Özgür, Ü., Alivov, Y.I., Liu, C., et al.: A comprehensive review of ZnO materials and devices. Applied Physics Reviews 98, 041301 (2005)
[7] Pan, Z.W., Dai, Z.R., Wang, Z.L.: Nanobelts of Semiconducting Oxides. Science 291, 1927–1949 (2001)
[8] Liu, B., Zeng, H.C.: Hydrothermal Synthesis of ZnO Nanorods in the Diameter Regime of 50 nm. Journal of American chemical society 125, 4430–4431 (2003)
[9] Qiu, Y.F., Yang, S.H.: ZnO Nanotetrapods: Controlled Vapor-Phase Synthesis and Application for Humidity Sensing. Adv. Funct. Mater. 17, 1345–1352 (2007)
[10] Traversa, E., Bearzotti, A.: A novel humidity-detection mechanism for ZnO dense pellets. Sensors and Actuators B 23, 181–186 (1995)
[11] Zhang, Y.S., Yu, K., Jiang, D.S., et al.: Zinc oxide nanorod and nanowire for humidity sensor. Applied Surface Science 242, 212–217 (2005)
[12] Zhou, X.F., Zhang, J., Jiang, T., et al.: Humidity detection by nanostructured ZnO: A wireless quartz crystal microbalance investigation. Sensors and Actuators A 137, 209–214 (2007)
[13] Traversa, E.: Ceramic sensors for humidity detection: the state-of-the-art and future developments. Sensors and Actuators B 23, 135–156 (1995)
[14] Barsan, N., Koziej, D., Weimar, U.: Metal oxide-based gas sensor research: How to? Sensors and Actuators B 121, 18–35 (2007)
[15] Feng, P., Wan, Q., Wang, T.H.: Contact-controlled sensing properties of flowerlike ZnO nanostructures. Applied Physics Letters 87, 213111 (2005)
[16] Zhang, J., Wang, S.R., Wang, Y., et al.: ZnO hollow spheres: Preparation, characterization, and gas sensing properties. Sensors and Actuators B: Chemical (2009)
[17] Rout, C.S., Raju, A.R., Govindaraj, A., et al.: Hydrogen sensors based on ZnO nanoparticles. Solid State Communications 138, 136–138 (2006)
[18] Vergatesan, S., Kima, H.-j., Cho, E.A., et al.: Journal of Power Sources 156, 294–299 (2006)
[19] Wang, Y., Yeow, J., Chen, L.-Y.: Synthesis of aligned zinc oxide nanorods for humidity sensing. In: 3rd International Conference on Sensing Technology, November 30- December 3, pp. 670–673 (2008)
[20] Bahanemann, D.W., Jormann, C., Hoffmann, M.R.: Journal of Physics Chemics 91, 3789–3798 (1987)
[21] Yawale, S.P., Yawale, S.S., Lamdhade, G.T.: Tin oxide and zinc oxide based doped humidity sensors. Sensors and Actuators A 135, 388–393 (2007)
[22] Gurlo, A., Riedel, R.: In Situ and Operando Spectroscopy for assessingmechanisms of gas sensing. Angew. Chem. Int. Ed. 46, 3826–3848 (2007)
[23] Williams, D.E.: Semiconducting oxides as gas-sensitive resistors. Sensors and Actuators B 57, 1–16 (1999)

Ultra-Wideband Radars for Through-Wall Imaging in Robotics

Jairo Alejandro Gomez and Graham Brooker

Australian Centre for Field Robotics (ACFR)
The University of Sydney
The Rose Street Building J04, NSW2006,
Australia
j.gomez@cas.edu.au
g.brooker@cas.edu.au

Abstract. In this chapter, a methodology to incorporate imaging techniques in distributed robotic applications using through-wall radars is presented. The main idea is to apply concepts from tomography and synthetic aperture radar to gather cross-sectional information of indoor environments using some recent developments in UWB technology. Most of these techniques have not been fully exploited within the robotics community although they were developed more than three decades ago.

1 Introduction

Having information about the internal appearance of a structure is important for a large number of disciplines including medicine [1, 2], dentistry, material and building inspection, quality assurance, security [3], search and rescue, intelligent surveillance and reconnaissance (ISR) [4], astronomy, archaeology, mining and seismic analysis among many others. This chapter aims to show how mobile robots with low frequency UWB radars can be used to reconstruct a cross-sectional slice of an indoor structure. The methodology presented here can be applied to other structures as long as they are not metallic.

If a building's cross section is considered as a 2D map, then it can be used for robot navigation. Since most sensors currently employed in mobile robots such as lasers and cameras only provide information about the environment's surfaces, if an indoor map is required then the robot must go inside the building to create it. However, if through-wall sensors are combined with inverse imaging techniques such as tomography or synthetic aperture radar (SAR), internal information would be available to robots during the data fusion stage, and then a richer map representation could be obtained [5, 6].

This enhanced map would be valuable for example during search and rescue operations in collapsed buildings after earthquakes or explosions where prior information is unknown or unreliable, and where producing a map while searching, for instance using SLAM, might be both risky and time consuming. Instead, having an indoor map created from the exterior before rescue personnel and robots enter would help navigation, victim location, and therefore improve the overall mission success.

From the perspective of intelligent surveillance and reconnaissance, the knowledge of internal static structure (e.g. the building cross section) can be used to improve the quality of the dynamic information produced by people and objects moving behind walls. This can be achieved using background suppression algorithms.

This chapter includes some concepts that will bring robotics and through-wall sensing technology together in distributed wireless sensor networks [7] [8] - firstly, UWB technology, specifically impulse radars [9, 10] and time modulated communications [11, 12, 13, 14] and secondly, inverse scattering imaging [2, 15, 16, 17, 18].

UWB radars are low cost, power efficient and light weight sensors compared to conventional radars, which make them attractive for wireless sensor network applications. Additionally, the large bandwidth available provides more information about the target and the environment they are seeing than narrowband radars do [9].

Section 3 shows that properly designed radars can be used for seeing through walls [4, 5]. This section provides a detailed description of a hybrid UWB radar with built-in communication capabilities currently being tested at the Australian Centre for Field Robotics (ACFR). When several of these radars are used to look at a given object from different displacements (angular or linear), the obtained information can be fused to get a cross section of the object itself [5]. Section 4 deals with all the data processing associated with this task [2]. Section 5 presents some final comments about challenges that need to be tackled in a multi-robotic radar imaging like this.

2 Background

2.1 Through-Wall Sensing

The capability of seeing through walls is important to many organizations including military and law enforcement [3]. The reality however, is that human vision cannot penetrate solid walls and therefore it is necessary to use a different sensor.

One option is to use X-ray based imagers but the feasibility of building compact systems that can be carried by a person / robot is uncertain and furthermore, health hazards due to x-ray radiation are a serious constraint.

A different option relies on RF signals. Our day-to-day experience shows that RF signals can propagate through walls, otherwise it would be impossible for us to receive or make calls inside a building using our mobile phones.

The selection of the RF frequency depends mainly on two factors, angular resolution and penetration capability. Fig. 1 shows the total attenuation for different materials typically found in walls as a function of the RF signal's frequency. Most non-metallic walls are fairly transparent to radar signals at low frequencies where as solid metallic walls are completely opaque. In fact, RF signals are completely blocked even by a thin aluminum foil such as those used

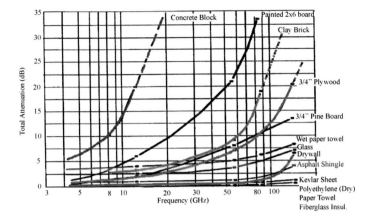

Fig. 1. RF attenuation caused by walls vs frequency. From [4].

in building insulation. In general, low-frequency radars have higher penetration capability but poorer angular resolution and vice versa. Radars suitable for through-wall sensing include UWB, X-band and millimeter-wave radars.

2.2 UWB

The Federal Communications Commission (FCC) of the United States defined in 2002 [19, 20, 21] an UWB transmitter as an intentional radiator that, at any point in time, has a fractional bandwidth $bw \geq 0.20$ given by (1), or has a UWB bandwidth ≥ 500 MHz, regardless of the fractional bandwidth. The UWB bandwidth was defined there as the frequency band bounded by the points with upper and lower frequencies located 10 dB below the highest radiated emission, as based on the complete transmission system including the antenna [14].

$$bw = 2\frac{f_H - f_L}{f_H + f_L} \qquad (1)$$

UWB communication and impulse radio schemes

Time-modulated ultra wideband (TM-UWB) communication is based on discontinuous emission of very short pulses [12], usually monocycles. The monocycle waveform can be any function which satisfies the spectral mask regulatory requirements (e.g. FCC regulation part 15, year 2002). Common pulse shapes include Gaussian, Laplacian, Rayleigh or Hermitian pulses.

TM-UWB does not require the use of additional carrier modulation because the pulse will propagate well in the radio channel. Therefore, this technique is baseband and it is referred as impulse radio (IR). In this case, one transmitted symbol is spread over N monocycles to achieve processing gain, similar than in spread spectrum systems, to combat noise and interference.

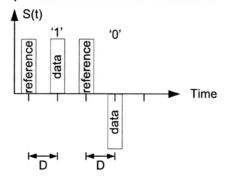

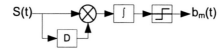

Fig. 2. a) Transmitted-reference and biphase modulation example. a) Non-coherent energy detector.

Because of the nature of impulse radio, simple methods can be used to encode information. Some modulation techniques include on-off keying, pulse-amplitude modulation, biphase modulation, pulse-position modulation and transmitted reference modulation [12].

Receivers for UWB communications can be either optimal receivers based on matched filters, where all paths need to be considered and are complex to implement, or sub-optimal receivers which are easier to design if transmitted reference modulation is used. In this case, the channel estimation is avoided through the use of non-coherent energy detectors [11]; see Fig. 2.

The UWB receiver in TM-UWB is a homodyne cross-correlator with a direct RF-to-baseband conversion [11]. Given the fact that intermediate frequency conversion is not needed, the implementation is simpler than in conventional (super-) heterodyne systems.

UWB transceiver

The proposed UWB transceiver design is composed of two main parts: an impulse range gated radar and a communication module. In this design, the transmitting stage is common to both subsystems, and the receiving section is special for each one of them. Fig. 3 shows a block diagram of the proposed design.

A square wave generator is used to feed a modulation stage, used mainly for communication purposes. The output of this stage is applied to a programmable delay line in the transmitter. This unit has two outputs, one of

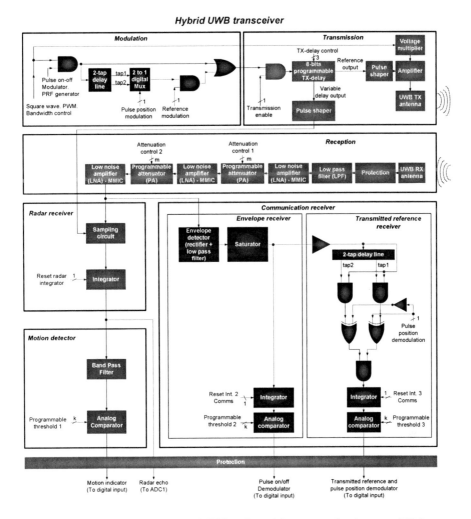

Fig. 3. Block diagram of the UWB radar with communication capabilities

them is the delayed signal used to control the trigger of the radar's time equivalent sampler, while the other output is used as input to a shaper block that helps to create short-time impulses that, after being amplified, are sent to the transmitting antenna. On the receiver side, the signal is received through the UWB antenna and it goes to a set of protection blocks, filters, low noise amplifiers and programmable attenuators. At this point, the signal is introduced to the radar receiver and communication receiver.

On the radar receiver side, the signal is digitalized using time equivalent sampling (that can be seen as a range gating technique) and applied to a band pass filter that is used as a motion detector. On the communication side of the transceiver, two demodulation methods are used. The first captures the envelope of the signal and after an integration process decides if the digit transmitted was '1' or '0'. The second method uses the advantages of the transmitted reference modulation mentioned in the previous section to recover the information from the incoming RF signal.

In this initial design, the multiple access capabilities required to handle the sensor network were not included in hardware. However, this is being implemented using time hopping. To keep the overall cost low, and enable this transceiver to be used in several robots, the central processing unit for every module is an ARM microcontroller.

2.3 Inverse Scattering Reconstruction Using Tomography

Tomography refers to the cross-sectional imaging of an object from either transmission or reflection data collected by illuminating the object from many different directions [1]. In Greek, "tomos" means section, or slice and "graphia" means to write or to draw.

The principle of tomography consists of measuring the spatial distribution of a physical quantity examining the sample under test from different directions and obtaining projections [2]. In strict sense of the word, a projection at a given angle is the integral of the image in the direction specified by that angle. However, in a general sense, a projection means the information derived from the transmitted energies, when an object is illuminated from a particular angle.

For performing tomographic imaging several sources can be used including X-rays, radio-isotopes, ultrasound, magnetic resonance and microwaves. This paper is concerned with microwave tomography which lies within the category of diffraction tomography [2] because the path followed by the emitted wave is subject to scattering and cannot be assumed to follow a straight line as in conventional X-ray tomography.

For diffracting sources, if the interaction between the object and the incident field is modeled with the wave equation, then a tomographic reconstruction approach based on the Fourier diffraction theorem is possible for weakly diffracting objects [16, 22]. Even though the ultimate interest lies in imaging the inhomogeneities within the object, there are no direct methods for solving the problem of propagation in an inhomogeneous medium. However, in practice, approximate formulations are derived from the theory of homogeneous medium wave propagation to create solutions in the presence of weak inhomogeneities. The better known among these methods are Born and Rytov approximations [23].

The wave equation is a nonlinear differential equation that relates an object to the surrounding fields. To estimate a cross-sectional image of an object $o(x, y)$, it is necessary to find a linear solution of the wave equation and then to invert this relation between the object and the scattered field. The necessary approximations for this purpose limit the range of objects that can be successfully imaged to those that do not severely change the incident field or have a small refractive index gradient compared to the surrounding media [2, 24].

3 Methodology

The following section presents a methodology for getting an image reconstruction of an object's cross section (e.g. a house) from the information provided by N randomly distributed robots equipped with low-frequency ultra-wideband (UWB) radars capable of seeing through walls.

Acquisition process

The intended process (see Fig. 4) can be described as follows:

1. Units are deployed around the object (see Fig. 4.1)
2. Units do a relative self-localization procedure using time of flight information (see Fig. 4.2).
3. A communication protocol is started to establish which unit is going to act as the active transmitter (beacon) while the rest act as receivers (see Fig. 4.3).
4. Once all the radar waveforms have been acquired in the receivers, a new unit is selected as the active transmitter and the process is repeated until the last unit has been used as beacon (see frames 4 to 10 in Fig. 4).
5. If there are insufficient units available to cover the object from all required angles, the units will have to move around the object and repeat the data collection process previously described in steps 1 to 4 (see Fig. 4.11).
6. Once the object has been seen from all the required angles, waveforms are processed to produce an image of the object's cross section (see Fig. 4.12) using the principles of inverse scattering theory. Of particular interest are theories from computerized tomographic imaging for diffracting sources in transmission mode [2] or from synthetic aperture radar imaging [16, 17].

Pre-processing and reconstruction

After collecting the data, it must be processed to be able to apply inverse imaging techniques. If a tomographic reconstruction for diffracting sources in transmission mode is considered [2] then the procedure is as follows:

1. Consider that the transmitter's temporal angular frequency w_0 [rad/s] is known or fixed from a set of frequencies available (considering that UWB signals are going to be used instead of narrowband signals).
2. The measurements of the N receivers will provide information about the total field in certain spatial positions (x, y) at a time t:

$$\mu(x, y, t) = \mu(x, y) e^{-jwt} \qquad (2)$$

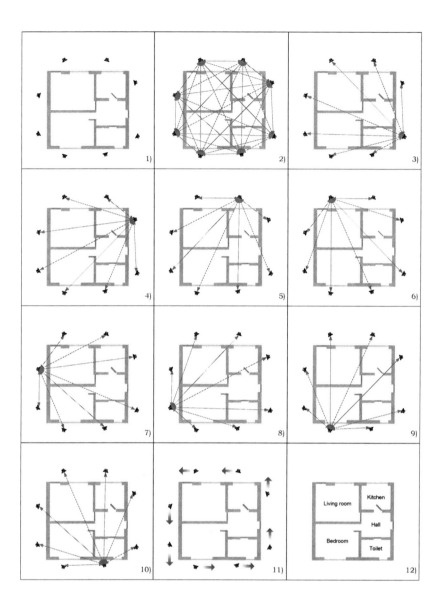

Fig. 4. Image reconstruction of an object's cross section from the information provided by N randomly distributed units equipped with low-frequency UWB radars

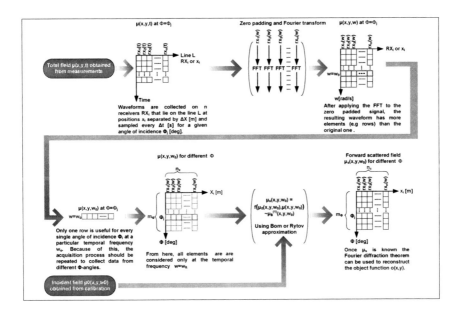

Fig. 5. Intermediate digital signal processing applied to the radar measurements

3. Applying the Fourier transform along the temporal component, and evaluating this for $w = w_0$, $\mu(x,y,w_0)$ is found:

$$\mu(x,y,w_0) = F\{\mu(x,y,t)\} \text{ at } w = w_0 \qquad (3)$$

4. The incident field on the receivers given by $\mu_0(x,y,w_0)$ and/or the incident complex phase given by $\Phi_0(x,y,w_0)$ are known (for example through calibration in a homogeneous medium).

5. Compute the forward scattered field $\mu_B^{(1)}(x,y,w_0)$ using Born or Rytov approximations.

   5.1 If the Born approximation is used:

$$\mu_s(x,y,w_0) = \mu(x,y,w_0) - \mu_0(x,y,w_0) \qquad (4)$$

$$\mu_B^{(1)}(x,y,w_0) \approx \mu_s(x,y,w_0) \qquad (5)$$

   5.2 If the Rytov approximation is used:

$$\Phi_s(x,y,w_0) = \Phi(x,y,w_0) - \Phi_0(x,y,w_0) \qquad (6)$$

$$\mu_B^{(1)}(x,y,w_0) \approx e^{\Phi_0(x,y,w_0)}\Phi_s(x,y,w_0) \qquad (7)$$

6. From $\mu_B^{(1)}(x,y,w_0)$ it is possible to recover the object function $o(x,y)$ using frequency interpolation or the filtered backpropagation algorithm from the Fourier diffraction theorem. Usually, the frequency interpolation is preferred over the filtered backpropagation algorithm because it is faster [2]. A complete review of these techniques for getting the function object or image cross-section from the scattered projections collected is found in [22].

Fig. 5 shows a detailed flow graph with all the DSP previously mentioned that is applied to the acquired data.

4 Conclusions

Expensive through-wall imaging equipment has limited the application of imaging techniques in robotics applications, but now UWB technology is promising to change that. It is possible to perform most of the steps required for the imaging using the proposed system with UWB transducers: self-localization, communication and synchronization, waveform generation and acquisition and finally image reconstruction using concepts from inverse scattering theory.

Most imaging techniques available require the object to be seen from different angles (tomography, ISAR) or from different positions for a given transducer aperture (SAR). This means that either you need to have a large number of units for doing it all at once, or you need to move the units you have around the object.

In tomography and synthetic aperture imaging a big assumption is that you know the position of your transducers relative to the object you want to image. This is not a problem in clinical applications where the environment is completely controlled, but it does impose some serious constraints in robotics applications where position uncertainty is always present.

The system we envisage would use a probabilistic framework for handling positioning errors and sensor noise, as well as a reconstruction method robust enough to be applied to big objects (something the Born assumption cannot handle) with large changes in the refractive index (where the Rytov assumption fails) surrounded by lossy medium and without the plane wave assumption. In general, the scalar approximation of the wave equation, keystone in the theory of standard diffraction tomography and SAR, will not be valid for heavily scattering objects. Instead a vectorial description for the inverse problem will be used and then more information will be obtained from polarization changes in the received wave due to scattering produced within object's inhomogeneities.

References

1. Kalender, W.A.: Computed tomography. Fundamentals, System technology, Image quality, applications, 2nd edn. Publicis, Munich (2005), www.imp.uni-erlangen.de
2. Slaney, A.C.K.M.: Principles of Computerized Tomography Imaging. Classics in applied mathematics. Society of Industrial and Applied Mathematics. Siam, Philadelphia (2001), http://www.slaney.org/pct
3. Baranowski, E.: Visibuilding: Sensing through walls. In: Fourth IEEE Workshop on Sensor Array and Multichannel Processing, July 2006, pp. 1–22 (2006)
4. Gauthier, S., Chamma, W.: Through-the-wall surveillance. Defence Research and Development Canada, DRDC Ottawa, Technical memorandum TM 2002-108 (October 2002)
5. Hunt, A.: Image formation through walls using a distributed radar sensor array. In: Proceedings of the 32nd Applied Imagery Pattern Recognition Workshop, October 2003, pp. 232–237 (2003)
6. Xue, R.-F., Yuan, B., Mao, J., Liu, Y.: Tomographic inverse scattering approach for radar imaging with multistatic acquisition. In: Proceedings of Geoscience and Remote Sensing Symposium, IGARSS 2005, July 2005, vol. 7, pp. 4623–4625 (2005)
7. Dutta, P.K., Arora, A.K.: Integrating micropower radar and motes. The Ohio State University, OSU-CISRC-12/03-TR67 (2003)
8. Gomez, J., Brooker, G.: Opportunities for imaging in distributed robotics applications with ultra-wideband radars. In: Proceedings of the third International Conference on Sensing Technology (ICST 2008), November 30-December 3, pp. 15–20 (2008)
9. Taylor, J.D. (ed.): Introduction to Ultra-Wideband Radar Systems. CRC, Boca Raton (1994)
10. Taylor, J.D. (ed.): Ultra-Wideband radar technology. CRC Press, Boca Raton (2000)
11. Nekoogar, F.: Ultra-Wideband Communications: Fundamentals and Applications. Prentice Hall, Englewood Cliffs (2005)
12. Oppermann, I., Hämäläinen, M., Iinatti, J.: UWB: Theory and Applications. Wiley, Chichester (2004)
13. Ghavami, M., Michael, L., Kohno, R.: Ultra Wideband Signals and Systems in Communication Engineering, 2nd edn. Wiley, Chichester (2007)
14. Schantz, H.: The Art and Science of Ultrawideband Antennas. Artech House, Boston (2005)
15. Skolnik, M.: Radar Handbook, 3rd edn. McGraw-Hill, New York (2008)
16. Nahamoo, D.: Ultrasonic diffraction imaging. Ph.D thesis, Purdue University (1982)
17. Soumekh, M.: Synthetic Aperture Radar Signal Processing with MATLAB Algorithms. Wiley-Interscience, New York (1999)
18. Cumming, I.G., Wong, F.H.: Digital Processing of Synthetic Aperture Radar Data: Algorithms And Implementation. Artech House Publishers, Norwood (2005)
19. FCC, Revision of part 15 of the commission's rules regarding ultra-wideband transmission systems, Tech. Rep. (2002)
20. FCC, Extract - executive summary of the fcc order of 4-22-2002, Tech. Rep. (2002)

21. Thomas, E.: Walk don't run: The first step in authorizing ultra-wideband technology. In: IEEE Conference on Ultra Wideband Systems and Technologies (UWBST) (May 2002)
22. Pan, S., Kak, A.: A computational study of reconstruction algorithms for diffraction tomography: Interpolation versus filtered-backpropagation. IEEE Transactions on Acoustics, Speech, and Signal Processing 31(5), 1262–1275 (1983)
23. Iwata, K., Nagata, R.: Calculation of refractive index distribution from interferograms using the born and rytov's approximation. Japanese Journal of Applied Physics 14, 379–384 (1975), http://jjap.ipap.jp/link?JJAPS/14S1/379
24. Slaney, M., Kak, A., Larsen, L.: Limitations of imaging with first-order diffraction tomography. IEEE Transactions on Microwave Theory and Techniques 32(8), 860–874 (1984)

Electrical and Gas Sensing Perfomanance of Coppergermanate

V.B. Gaikwad[1], A.V. Borhade[2], Y.R. Baste[3], D.D. Kajale[3], and G.H. Jain[3]

[1] K.T.H.M. College, Nashik 422005, India
[2] H.P.T. Arts & R.Y.K. Science College, Nashik 422005, India
[3] Arts, Commerce and Science College, Nandgaon, Nashik 423106, India

Abstract. This study reports the preparation and gas-sensing performanance of thick film coppergermanate (CuGeO$_3$) based gas sensor. The structural and microstructural properties of CuGeO3 material have been characterized using by various analytical practice. Thick films of the material were prepared by screen printing technique. The electrical resistance response of the sensor was investigated for different gases like H$_2$S, LPG, CO, CO$_2$, O$_2$ and NH$_3$ at different operating temperatures ranging from room temperature to 450ºC. Selectivity and sensitivity of CuGeO$_3$ is measured and reported. The material shows negative temperature coefficient. The synthesized sensor shows excellent electrical resistance response toward H$_2$S gas.

Keywords: Gas sensor, Thick film, Selectivity, Sensitivity, Response time.

1 Introduction

Hazardous and toxic gases from auto and industrial exhausts are polluting the environment. The detection of toxic and flammable gases is a matter of urgent public concern. The need to detect, measure and control these gases has led to the research and development of a wide variety of sensors using different materials and technology [1-2]. Several efforts have been made in this direction over the entire world during last five decade. It mainly includes the search for and development of new material that will give promising result for gas sensing characteristics. However, percentage production of new gas sensing material still lags behind actual requirement. Chemical sensors have the potential to play an influential role in meeting environmental objectives through application in several sectors. These roles would take advantage of number of favorable attributes of chemical sensors, such as their relatively low cost, ruggedness and capability for miniaturization. The gas sensing properties of presently, investigated materials are found to be continuously increasing by simply changing various parameters, like additive concentration, firing temperature, synthesis method, processing condition and development criteria in the devices. This gives an innovative direction for the newer and newer materials for gas sensing application.

Chemical sensors must provide sufficient sensitivity for required measurement adequate long-term reliability at all competitive cost. Semiconductor sensors based on this property have been used extensively for chemical and gas detection, which included toxic and polluted gases. (H_2S, CO, CO_2, NH_3, NO_x, Cl_2 and SO_2) and combustible gases (H_2, CH_4, and flammable organic vapors) [3]. The requirement for the detection of toxic and flammable gases has led to develop of gas sensing device [4].

It is well known that the perovskite type oxides show a high electrical conductivity and exhibit oxidation-reduction catalytic characteristics [5-6]. Numbers of pervoskite oxide (ABO_3) have been used as gas sensor because of stability in thermal and chemical atmosphere. The pervoskite oxide ceramics have been created and promoted interest in new chemical sensors. The gas sensing pervoskite oxide ceramic could be classified into two groups: semiconducting ceramics and electrochemical ceramics. Semiconducting pervoskite oxide ceramics were widely used detection of reducing gases (such as CO [7-9], H_2 [10], C_2H_5OH [11]), oxygenic gases (CO_2 [12], O_2 [13-14], NOx [15],) and toxic gases NO_2 [16]. Hydrogen sulphide a toxic gas is often produced in coal, coal oil or natural gas industries [17]. The threshold-limited value (TLV) of H_2S is 10 ppm. When concentration of H_2S is higher than 250 ppm are likely to result in neurobehavioral toxicity and may cause death [18]. Monitoring of H_2S gas is crucial in laboratories and industrial areas. There are several compounds, which are reported as H_2S gas sensor [19-25].

2 Preparation of the Sensor Material

The stiochiometric mixture of CuO and GeO_2 (AR Grade) was subjected to stepwise calcination until terminal temperature. For the temperature, reaction mixture was heated in the muffle furnace with increase in temperature at the rate of 10°C per minute from one temperature to the subsequent higher temperature. After heating at higher temperature the material was cooled and grounded with gap of three hour using mortar and pestle. Later on, the ground material was further heated at 800 to 1100 °C for 12 hours. The material again grounded with gap of three hour using mortar and pestle. The product obtained used for analysis.

The thixotropic paste was formulated by mixing the fine powder of $CuGeO_3$ with ethyl cellulose (a temporary binder) in a mixture of organic solvent. The ratio of inorganic to organic part was kept at 75:25 in formulating the paste. This paste was screen printed on glass substrate in the desired pattern [26-27]. To remove organic binder the thick films were fired at 550°C. The silver paste was used for ohmic contacts.

3 Static Gas Sensing Unit

The gas sensing performance of the sensor was examined using a 'static gas sensing system' shown in figure1.There were electrical feeds through the base plate. The heater was fixed on the base plate to heat the sample under test up to required operating temperatures. The current passing through the heating element was monitored using a relay operated with an electronic circuit with adjustable ON-OFF time intervals. A Cr-Al thermocouple was used to sense the operating temperature of the sensor. The output of the thermocouple was connected to a digital temperature indicator. A gas inlet valve was fitted at one of the ports of the base plate. The required gas concentration inside the static system was achieved by injecting a known volume of a test gas using a gas-injecting syringe. A constant voltage was applied to the sensor, and the current was measured by a digital picoammeter. The air was allowed to pass into the glass chamber after every gas exposure cycle.

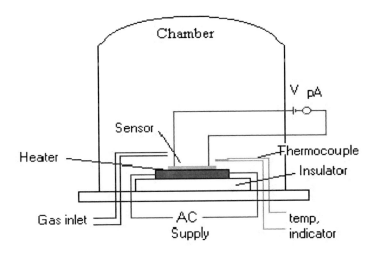

Fig. 1. Static gas sensing unit

4 Characterization

The structural properties of the powder were studied using various analytical techniques. X-ray diffractogram studied for diffraction of X-rays from planes of the crystal and plane spacing with Rinku model DMAX-2500 X-ray diffractometer (XRD) with Cu-Kα radiation, having $\lambda = 1.5406$Å. The surface morphology of the synthesized material was analyzed using scanning electron microscope (SEM, JEOL JED 2300LA). Absorption band of functional group in material was studied by Fourier Transform Infrared Spectrometer (FTIR) using Spectrum-2000 FTIR spectrophotometer in the range of 450-4000cm^{-1}.

4.1 Structural Properties

X-ray diffractogram of a thick film is shown in figure 2. It revealed from XRD that the material is polycrystalline in nature with orthorhombic pervoskite phase. The observed peak positions are good agreement with the reported value in JCPDS data card number 74-0302 confirming orthorhombic phase. The average grain size was determined by using Scherer formula and was estimated to be 49

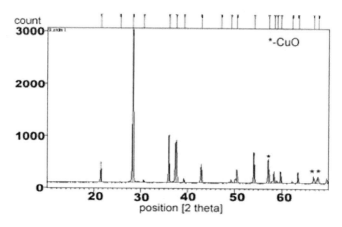

Fig. 2. XRD of CuGeO$_3$

nm. Few peaks of CuO were observed at the diffraction angle of 66.72° and 67.72°. This indicates that presence of trace amount of CuO in synthesized sensor, which matters H$_2$S sensing performance.

4.2 Microstructural Analysis

Figure 3 illustrate the scanning electron micrograph (SEM) image of pure CuGeO$_3$ thick film at different magnification. The film consist of large number of particle size ranging from 1μm to 10μm, leading to high porosity and large effective surface area which is available for the adsorption. The film is observed to be porous and particles are uniformly distributed. Due to high degree of porosity of the film leading to small surface to volume ratio.

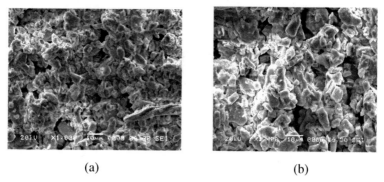

(a)   (b)

Fig. 3. (a and b) SEM of CuGeO$_3$ at different magnification

4.3 FTIR Spectra

Figure 4 shows FTIR spectra of $CuGeO_3$. The spectrum consists of peak at 850cm^{-1} indicates the presence of Ge-0 vibration frequency in $CuGreO_3$. The spectra also show vibration frequency at 630, 1620 and 1450 cm^{-1} clearly indicates presence of Cu-O- Cu in $CuGeO_3$. This confirms the presence of trace amount of CuO in the synthesized sensor material.

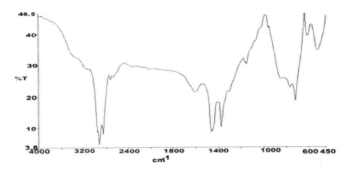

Fig. 4. FTIR spectra of $CuGeO_3$

5 Electrical Properties

Electrical performance of the material was studied by measuring change in conductance with temperature. Figure 5 shows the dependence of conductivity of coppergermanate film in air ambience. The conductivity of the film goes on increasing with increase in temperature, indicating negative temperature coefficient

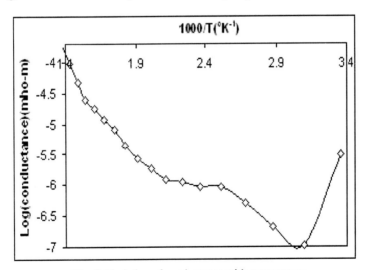

Fig. 5. Variation of conductance with temperature

(NTC) of the resistance. This shows the semiconducting nature of the film. The material showed positive temperature below 323°K as per the phase transition of the pevoskiteoxides [28].

5.1 Gas Sensing Performance

Gas Response (S) is defined as the ratio of change in conductance of the sample on exposure to gas to the conductance in air.

$$S = \frac{Gg - Ga}{Ga} = \frac{\Delta G}{Ga}$$

Where Gg and Ga are the conductance of the sample in presence and in absence of the test gas respectively, and ΔG is change in conductance.

The synthesized sensor was tested for various gases like NH_3, H_2, LPG, O_2, H_2S, CO and CO_2 at various temperatures ranging from room temperature to 450°C. Figure 6 represents the variation in the sensitivity to H_2S gas (10ppm) with operating temperature. It is noted from the graph that response increases with increasing temperature, and attains a maximum at 50°C, and decreases with further increase in operating temperature (S = 215).

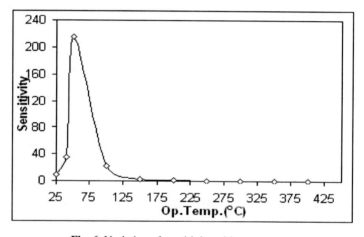

Fig. 6. Variation of sensitivity with temperature

5.2 Selectivity of CuGeO₃ Film

Selectivity is the ability of sensor to respond to certain gas in presence of other gases. The time taken for the sensor to attain 90% of the maximum change in resistance on exposure to the gas is the response time. The time taken for the sensor to get back 90% of the original resistance is the recovery time. Figure 7 presents the bar diagram indicating selectivity of Coppergermanate film at 300°C to H_2S gas against the other gases. The sensor showed poor selectivity, which is the conventional characteristic of metal semiconductor oxide. Efforts were made to detect the gas sensing performance at lower temperature. The sensor showed tremendous response (215) to H$_2$S gas for dilute concentration (10ppm) as shown in figure 5. This was the unique feature of the synthesized material.

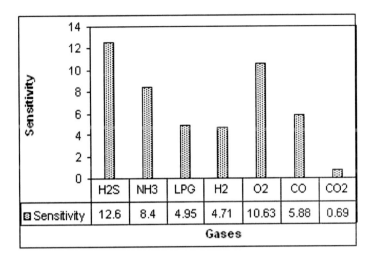

Fig. 7. Selectivity of CuGeO3 films at 300°C

5.3 Response and Recovery Time

The response and recovery time of CuGeO$_3$ film was found to be 6s and 180s respectively. This can be explained in following discussion.

6 Discussion

The XRD of the material consist of a few peaks of CuO at an angle of diffraction 66.72° and 67.72°. The CuO is a p-type semiconductor. The coppergermanate pervoskite oxide was found to be of n-type. Therefore the film can be looked upon consist of p-CuO and n-CuGeO$_3$ oxide.

The mechanism to explain the very large decrease in resistance of thin films on exposure to H$_2$S gas was suggested in the literature [29-30]. CuO and CuGeO$_3$ which are p and n-type semiconductors respectively, have a strong electronic interaction due to which the CuO– CuGeO$_3$ surface consists of numerous p–n junctions causing very high resistance of films in air. On exposure to H$_2$S gas, CuO particles are rapidly converted to CuS by the following chemical reaction:

$$CuO + H_2S \rightarrow CuS + H_2O \qquad (1)$$

CuS is metallic in nature and its formation destroys the p–n junctions existing on the surface causing a large decrease in electrical resistance. On the other hand, CuS will be oxidized in air and will change back to CuO reversibly through the reaction as follows:

$$CuS + 3/2O_2 \rightarrow CuO + SO_2 \qquad (2)$$

In fact, these two reactive processes dominate the response and recovery of the sensor. This explains why the recovery is poor at low temperature. Since the rate of oxidation increases with temperature, recovery time decreases with increase in operating temperature. However, increasing sensor operating temperature disrupts the p–n junctions even in air leading to a fall in electrical resistance and hence its sensitivity to H_2S. The unstable CuS at a high temperature is responsible for the decrease in sensitivity when raises to 215.

7 Summary

The coppergermanate was synthesized by solid-state reaction method. The synthesized material sensor possesses the negative temperature coefficient. Thick films showed weak selectivity at higher temperatures. It showed response and good selectivity to H_2S gas at lower temperature (50°C) even at room temperature for the dilute gas concentration (10ppm). The gas detection of coppergermanate thick films is based upon barrier potential, adsorption and absorption of oxygen species

References

[1] Arakawa, T., Kurachi, H., Shiokawa, J.: Physiochemical properties of rare earth pervoskite oxides used as gas sensor material. Journal of Material Science 20, 1207–1210 (1985)
[2] Moseley, P.T.: Materials selection for semiconductor gas sensors. Sensors and Actuators B 6, 149–156 (1992)
[3] Noval, T.G., Yordanav, S.P.: Ceramic sensors- Technology and Application. Technomic Publishing Lancaster, PA (1996)
[4] Rue, G.H., Ban, T.H., Choi, N.J., Kwak, J.H., Lim, Y.T., Lee, D.D.: Toxic gas response of (In,Sn)O2/Pt nanowire sensors. Solid-State Sensors, Actuators and Microsystems. Digest of Technical Papers 2(5-9), 1899–1902 (2005)
[5] Voorhoeve, R.J.H., Remeika, J.P., Freeland, P.E.: Rare-Earth Oxides of Manganese and Cobalt Rival Platinum for the Treatment of Carbon Monoxide in Auto Exhaust. Science 177(4046), 353–354 (1972)
[6] Gallagher, P.K., Jonson, D.W.: J. Amer. Ceram. Soc. 60, 128 (1977)
[7] Kuwabara, M., Ide, T.: CO gas sensitivity in porous semiconducting barium titanate. Amre. Ceram. Soc. Bull. 66, 1401–1405 (1987)
[8] Chiu, C.M., Chang, Y.H.: The structure, electrical and sensing properties for CO of the La0.8 Sr0.2CO 1-x NixO3 system. Mater. Sci. Eng., A 266, 93–98 (1999)
[9] Sorita, R., Kawano, T.: A highly selective CO sensor: Screening of electrode materials. Sensors and Actuators. B 35/36, 274–277 (1996)
[10] Inaba, T., Saji, K., Takahashi, H.: Limiting current-type gas sensor using high temperature -type conductor thin film. Electrochemistry 67, 458–462 (1999)
[11] Kong, L.B., Sen, Y.S.: Gas sensing property and mechanism of CaxLa1-xFe2O3. Sensors and Actuators. B 30, 217–221 (1996)
[12] Ishihara, N.Y., Nishiguchi, H., Takita, Y.: Detection mechanism of CuO-BaTiO3 capacitive type gas sensor. In: Proceedings of Ceramic Sensors III, San Antonio, TX, USA, pp. 6–11 (1997)

[13] Noguchi, Y., Kuroiwa, H., Takata, M.: Sensing properties of oxygen sensor using hot spot on La0.8Sr0.8Co0.8Fe0.2O3. Ceramic rod, Key Eng. Mater., 160–170, 79–82 (1999)
[14] Lukaszewicz, J.P., Miura, N., Yamazoe, N.: A LaF3 based oxygen sensor with pervoskite-type oxide electrode operative at a room temperature. Sens. and Actuat. B 1, 95–198 (1990)
[15] Shuk, P., Kharton, V., Tichonova, L., Wiemhofer, H.D., Guth, U., Gopel, W.: Electrodes for oxygen sensors based on rare earth magnates or cobaltite. Sensors and Actuators ctuat. B 16, 401–405 (1993)
[16] Yamaura, H., Tamaki, J., Miura, N., Yamazoe, N.: NOx sensing properties of metal titanate based semiconductor sensor at elevated temperature, Engineering Science Reports, Kyushu University, Japan, pp. 341–346 (1995)
[17] Dorman, D.C., Brenneman, K.A., Struve, M.F., Miller, J.R.A., Marshal, M.W.: Fertility and developmental neurotoxicity effects of inhaled hydrogen sulfide in Sprague-Dawley rats. Neurotoxicology and Teratology 22(1), 71–84 (2000)
[18] Struve, M.F., Brisbois, J.N., James, R.A., Marshal, M.W., Dorman, D.C.: Nurotoxicology 22, 375C (2001)
[19] Jin, T., Yamazaki, I.K., Kikuta, T., Nakatani, N.: H2S sensing property of porous SnO2 sputtered films coated with various doping films. Vacuum 80, 723–725 (2006)
[20] Jain, G.H., Patil, L.A.: CuO-doped BSST thick film resistors for ppb level H2S gas sensing at room temperature. Sensors and Actuators. B 123, 246–253 (2007)
[21] Sen, S., Bhandarkar, V., Muthea, K.P., Roy, M., Deshpande, S.K., Aiyer, R.C., Gupta, S.K., Yakhmi, J.V., Sahni, V.C.: Highly sensitive hydrogen sulphide sensors operable at room temperature. Sensors and Actuators B 115, 270–275 (2006)
[22] Malyshev, V.V., Pislyakov, A.V.: SnO2-based thick-film-resistive sensor for H2S detection in the concentration range of 1–10 mg m-3. Sensors and Actuators B 47, 181–188 (1998)
[23] Frtihberger, B., Grunze, M., Dwyer, D.J.: Surface chemistry of H2S sensitive tungsten oxide films. Sensors and Actuator; B 31, 167–174 (1996)
[24] Ianghua, K., Yadong, L.: High sensitivity of CuO modified SnO2 nanoribbons to H2S at room temperature. Sensors and Actuators B 105, 449–453 (2005)
[25] Niranjan, R.S., Patil, K.R., Sainkar, S.R., Mulla, I.S.: High H2S-sensitive copper-doped tin oxide thin film. Materials Chemistry and Physics 80, 250–256 (2003)
[26] Wagh, M.S., Jain, G.H., Patil, D.R., Patil, S.A., Patil, L.A.: Modified zinc oxide thick films resistors as NH3 gas sensors. Sensors and Actuators B 115, 125–133 (2006)
[27] Jain, G.H., Patil, L.A., Wagh, M.S., Patil, D.R., Patil, S.A., Amalnerkar, D.P.: Surface modified BaTiO3 thick film resistors as H2S gas sensor. Sensors and Actuators B 117, 159–165 (2006)
[28] Zhou, Z.G., Tang, Z.L., Zhang, Z.T., Wlodarski, W.: Pervoskite oxides of PTCR ceramics as a chemical sensors. Sens. Actuators B 77, 22–26 (2001)
[29] Devi, G.S., Manorama, S., Rao, V.J.: High sensitivity and selectivity of an SnO2 to H2S at around 100 oC. Sensors and Actuators, B 28, 31–37 (1995)
[30] Chowdhuri, A., Sharma, P., Gupta, V., Sreenivas, K., Rao, K.: H2S gas sensing mechanism of SnO2 films with ultrathin CuO dotted islands. Journal of applied physics, 2172–2180 (2002)

Studies on Gas Sensing Performance of Pure and Nano- Ag Doped ZnO Thick Film Resistors

V.B. Gaikwad[1], M.K. Deore[2], P.K. Khanna[3], D.D. Kajale[2], S.D. Shinde[1], D.N. Chavan[1], and G.H. Jain[2,*]

[1] K.T.H.M. College, Nashik 422005, India
[2] Dept. of Physics, Arts, Commerce and Science College
  Nandgaon 423106, India
[3] C-MET, Pune. 411008, India
*Corresponding author : gotanjain@rediffmail.com

Abstract. Thick films of AR grade ZnO were prepared by screen-printing technique. The gas sensing performances of thick films were tested for various gases. It showed maximum sensitivity to CO gas at 100 °C for 100ppm gas concentration. To improve the sensitivity and selectivity of the film towards a particular gas, ZnO thick films were surface modified by dipping them in a solution of nano silver for different intervals of time. These surface modified ZnO films showed larger sensitivity to H_2S gas (100ppm) than pure ZnO film at 300 °C. Nano silver on the surface of the film shifts the reactivity of film from CO to H_2S gas. A systematic study of sensing performance of the sensor indicates the key role-played by the nano silver species on the surface.The sensitivity, selectivity, response and recovery time of the sensor were measured and presented.

Keywords: ZnO gas sensor, thick films, nano silver, sensitivity, selectivity, CO gas sensor, H_2S gas sensor.

1 Introduction

Semiconductor gases sensors based on metal oxides have been used extensively to detect toxic and inflammable gases [1, 2].Metal oxides such as SnO_2, Sb_2O_3, Fe_2O_3, ZnO, and Fe_2O_3 etc. are potential candidates for developing portable and inexpensive gas sensing devices, which have the advantages of simplicity, high sensitivity and fast response. In recent years, the concern over environmental protection and increasing demands to monitor hazardous gases in industries and home has attracted extensive interest in developing gas sensors for various polluting and toxic gases such as CO, CO_2, NOx, and H_2S. Zinc oxide is one of the earliest discovered metal oxide gas sensing material. It is an n-type semiconductor of wurtzite structure with direct energy wide band gap of about 3.37eV at room temperature [3], and very large excitation energy of about 60MeV. ZnO is an important oxide semiconductor for sensing applications to toxic and combustible gases [4, 5]. Generally, ZnO sensor provided a wide variety of advantages, such as low cost, short response time, easy manufacturing and small size, compared to traditional analytical instruments. However its working temperature is rather high, normally at 400-500°C, and selectivity response ability is fairly poor, in recent years, the study on ZnO gas sensing material has become one of the major research topics, and the research is focused on improving their preparation method and decreasing their working temperature [6, 8].

In general, metal oxide materials are rarely used as a single phase for sensors. And their gas sensing characteristics are usually well modified by adding a small amount of catalysts. Platinum and Palladium are well-known active catalysts, which enhance the sensitivity against reducing gases [9]. Gold (Au), Silver (Ag) are also utilized to enhance the sensitivity of oxide materials [10]. It is believed that the catalysts promote chemisorptions process and thus increases the density of chemisorbed oxygen species, which are reaction centres for reducing gas molecules. The addition of second component in metal oxide semiconductor gas sensor either as bulk doping or as surface modification is one of the successful method to optimize and improve the properties of a gas sensors [11-16].

In the present study, the nano-Ag is used for modification of ZnO film. It has been described that the nano Ag particles on the surface of ZnO films by dipping technique improves the surface adsorption capability. As a result of the numerous interfaces between the nano-grains that interact with gas molecules, can effectively improve the diffusion properties, while greatly increasing the gas sensitivity of the materials [14-16]. The nano Ag used in the present work was prepared as described in the literature [17].

In this work, ZnO thick films were prepared and deposited by screen printing on to glass substrates .The film surface was modified by dipping them into the solution of nano Ag. The sensing performance of the film was tested by static gas sensing system for different gases such as H_2S, CO, ethanol etc.

2 Operating Principle of Gas Sensor

Since long it has been known that adsorption of reducing gas molecules results in decrease in electrical resistance of oxide material [18]. The ZnO materials are characteristically n-type semiconductor due to non-stoichometry associated with oxygen vacancy and/or metal excess which acts as donor states thus providing conduction electrons. However, the overall surface resistance of such films is generally influenced by chemisorptions (chemical adsorption) of oxygen from air on the surface and at the grain boundaries. The chemisorbed oxygen traps conduction electrons and remains as negatively charged species (O_2^-, O^- or O^{2-} depending on temperature.) on the surface [19]. The process results in an increase of surface resistance. In presence of reducing gases the trapped electrons are released due to the reaction between the gas molecules and negatively charged chemisorbed oxygen species resulting in decreasing in resistance of the materials.

When the gas is removed from the sensor environment, the resistance again increases and the material recovered to original resistance.

3 Experimental

3.1 Preparation of Pure and Modified ZnO Thick Films

The AR grade powder ZnO was milled for 2h to obtain fine-grained powder. The thixotropic paste was formulated by mixing the fine powder of ZnO with a solution of ethyl cellulose (a temporary binder) in a mixture of organic solvents such as butyl cellulose, butyl carbitol acetate and terpineol, etc. The ratio of the inorganic to organic part was kept at 75:25 in formulating the paste. This paste was screen printed on a glass substrate in a desired pattern [20, 21]. The films were fired at 550 °C for 30 min. Silver contacts are made for electrical measurements. The pure ZnO films were surface modified by dipping them in to 2% nano silver solution for different intervals of time: 5, 10, 20, and 30min as explained elsewhere [22]. These films were dried at 80°C, followed by firing at 550°C for 30 minutes.

3.2 Details of Gas Sensing Unit

The sensing performance of the sensors was examined using a 'static gas sensing system'. There were electrical feeds through the base plate. The heater was fixed on the base plate to heat the sample under test up to required operating temperatures. The current passing through the heating element was monitored using a relay operated with an electronic circuit with adjustable ON-OFF time intervals. A Cr-Al thermocouple was used to sense the operating temperature of the sensor. The output of the thermocouple was connected to a digital temperature indicator. A gas inlet valve was fitted at one of the ports of the base plate. The required gas concentration inside the static system was achieved by injecting a known volume of a test gas using a gas-injecting syringe. A constant voltage was applied to the sensor, and a digital Pico ammeter measured the current. The air was allowed to pass into the glass chamber after every gas exposure cycle.

4 Characterizations

4.1 Structural Analysis

The crystalline structure of the films was analyzed with X-ray diffractogram using CuKα radiation with wavelength 1.5418 Å. Figure 1, depicts the XRD patterns of pure ZnO. The observed peaks in figure are matched well with ASTM reported data of ZnO [23].

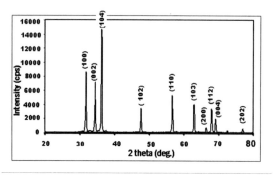

Fig. 1. X-ray diffractogram of pure ZnO

4.2 Micro Structural Analysis

Figure 2, show the micrographs of pure ZnO and surface modified ZnO. The comparison of these micrographs shows the interesting changes in morphology. Figure 2 (a), consists of large number of grain sizes ranging from 0.1μm to 1μm,

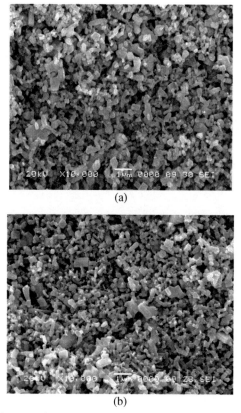

Fig. 2. SEM images of (a) pure ZnO and (b) modified ZnO (10min.) films

leading to high porosity and large effective surface area available for the adsorption of oxygen species. Figure 2 (b), shows a number of small particles distributed uniformly between the larger grains around the ZnO, which may be attributed to the presence of nano Ag. The grain size range was observed to be from 0.05μm to 0.9μm. The presence of nano-Ag particles on the surface of the film alters the adsorption-desorption ability of the film. The altered surface topography shifts the affinity and reactivity of film towards the H_2S gas. EDAX analysis of silver modified film shows the presence of silver and the appearance from the film indicates that the silver particles are present at the surface of the ZnO film.

4.3 Electrical Properties

4.3.1 I-V Characteristics

Figure 3, shows the I-V characteristics of pure ZnO film at room temperature. I-V Characteristics are observed to be symmetrical in nature, indicating the ohmic nature of silver contact.

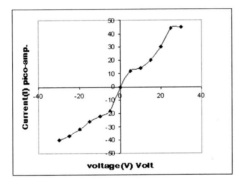

Fig. 3. I-V Characteristics of pure ZnO

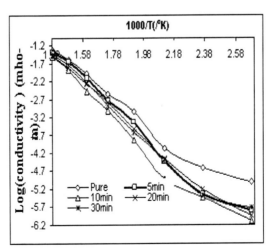

Fig. 4. Variation of conductivity with temperature

4.3.2 Electrical Conductivity

Figure 4, shows the dependence of conductivity of pure ZnO and surface modified ZnO films in air ambience. The conductivity of these films goes on increasing with increase in temperature, indicating negative temperature coefficient (NTC) of resistance. This shows the semi conducting nature of the films.

5 Gas Sensing Performances

5.1 Gas Sensing Performance of Pure ZnO Film

5.1.1 Gas Response with Operating Temperature

Gas response (S) of gas sensor is defined as the ratio of change in conductance of the sensor on exposure to the target gas to the original conductance in air [24].

$$S = (Gg - Ga) / Ga$$

Where   Ga= conductance of sensor in air.
        Gg = Conductance of sensor in gas.

Figure 5, shows the variation in the gas response of CO gas (100ppm) with operating temperatures ranging from 50 °C. to 450 °C. It is noted from the graph that response increases with increasing temperature, and attains a maximum at 100 °C, and decreases with further increase in operating temperature.

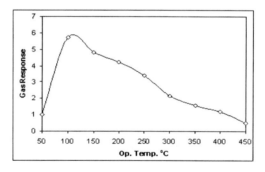

Fig. 5. Variation of response with temperature

5.1.2 Selectivity of Pure ZnO Film

Selectivity is defined as defined as the ability of sensor to respond to a certain gas in the presence of other gases. Selectivity is another important parameter of a gas sensor .The sensor must have rather high selectivity for its application. ZnO film is examined for different gases at different operating temperatures and the results are shown in figure 6.The bar diagram indicating selectivity of pure ZnO film at 100°C to CO gas against the other gases. The sensor is the most selective to CO gas against the other gases except H_2S.

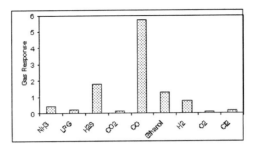

Fig. 6. Selectivity of pure ZnO film

5.1.3 Response and Recovery Time of Pure ZnO Film

Response and recovery time are the basic parameters of the gas sensors. Which are defined as the time taken for the sensor to attain 90% of maximum change in resistance on exposure to gas is the response time. The time taken by the sensor to get back 90% of the original resistance is the recovery time [25]. The response and recovery time of pure ZnO film was 4s and 60s respectively. The large recovery time would be due to lower operating temperature. At lower temperature O2- species is more prominently adsorbed on the surface and thus it is less reactive compared to other species of oxygen, O^- and O^{--}.

5.2 Gas Sensing Performance of Modified ZnO Films

The enhanced response of modified sensor can be attributed to two factors. First is higher specific area of the modified sensor can lead to increase in active surface for gas sensing and second, the catalyst enhances the adsorption of gas molecules and accelerates the electron exchange between the sensor and test gas [26]. The two factors together contribute to the improvement of gas sensing properties of surface modified ZnO sensor. The pure ZnO films were surface modified by dipping them in to 2% nano silver solution for different intervals of time: 5, 10, 20, and 30min. These films were dried at 80°C, followed by firing at 550°C for 30min.

5.2.1 Gas Response with Operating Temperature

Figure 7, shows the variation in the gas response of pure and surface modified ZnO films to H_2S gas (100ppm) with operating temperatures ranging from 50°C to 450°C. It is noted from the graph that response increases with dipping time attains maximum for 10min film and decreases on further increase in dipping time. The pure film showed highest response to CO, while modified films shows to H_2S gas. The alteration in nature of gas sensing from Co to H_2S for the modified ZnO sensor could be largely due to possibility of formation of silver sulpphide more easily than formation of silver oxide.

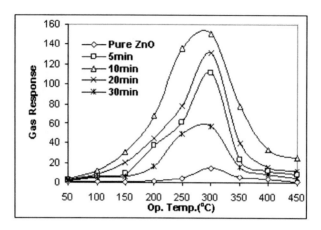

Fig. 7. Variation of gas response with temperature

5.2.2 Selectivity of Modified ZnO Film

Figure 8, represents the bar diagram indicating selectivity of surface modified ZnO film dipped for 10min at 300 °C to H_2S gas against the other gases. The sensor is the most selective to H_2S gas.

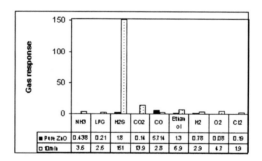

Fig. 8. Selectivity of pure and modified ZnO films

5.2.3 Response and Recovery Time of Modified ZnO Film

The response and recovery time of modified ZnO (10min) film was 3s and 90s respectively. The recovery time was found to be larger compared to pure ZnO film.

6 Effect of Dipping Time and the Amount of Nano Ag

Figure 9, shows the variation of gas response with dipping time and the amount of nano Ag. The film dipped for 10 min. showed the maximum gas response. At 10min.dipping time, the senor would find Ag mass %(0.44) to be optimum. The optimum mass % of Ag (0.44) would cover the film surface uniformly. The lower

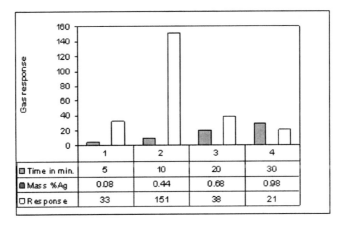

Fig. 9. Variation of response with dipping time the amount of nano Ag

gas response at higher operating temperature would be due to consumption of H_2S gas by adsorbed oxygen.

Figure 9, also represents the variation of H_2S gas response with the mass % of Ag (0.44) on ZnO film after surface modification. It is observed that the gas response is maximum at mass % of Ag (0.44). At higher mass%; the Ag surfactant would mask the base material resist the gas to reach to the surface site. So that gas response decreases.

7 Discussions

7.1 Pure ZnO as CO Gas Sensor

The gas sensing mechanism of the ZnO based sensors belongs to the surface controlled type, which is based on the change in conductance of the semiconductor. The oxygen adsorbed on the surface directly influence the conductance of ZnO based sensors. The amount of oxygen adsorbed on the sensor surface depends on operating temperature, partial size and specific surface area of the sensor [27]. The atmospheric oxygen molecules adsorbed on the surface of n-type ZnO in the form of O, O_2^- and O^{2-}, there by decreasing the conductance. The state of oxygen on the surface of ZnO sensor undergoing the following reactions [28].

$$O_2 \text{ (gas)} \rightarrow O_2 \text{ (ads)} \quad (1)$$

$$O_2 \text{ (ads)} + e^- \rightarrow O_2^- \text{ (ads)} \quad (2)$$

$$O_2^- \text{ (ads)} + e^- \rightarrow 2O^- \text{ (ads)} \quad (3)$$

$$O^- \text{ (ads)} + e^- \rightarrow O^{2-} \text{ (ads)} \quad (4)$$

The oxygen species captures electron from the material, which results in the concentration changes of holes or electrons in ZnO semiconductor.

Figure 10, shows the adsorption of oxygen species on the surface of Zink oxide, abstracting electrons and thus causing an increase in potential barrier of the grain boundaries.

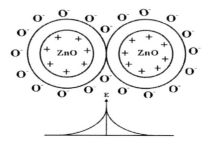

Fig. 10. Oxygen adsorption on surface of pure ZnO

Figure11, shows when ZnO comes in contact with CO gas the potential barrier would decrease as result oxidative conversion of CO gas and desorption of oxygen.

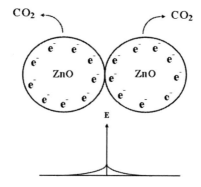

Fig. 11. Oxygen desorption on surface of pure ZnO

When ZnO films are exposed to the reducing gases, the reductive gas reacts with the oxygen adsorbed on the surface. Then the electrons are released back to the semiconductor, resulting in the change tin the electrical conductance of ZnO sensor. On the surface of the ZnO film there are two kinds of oxygen, one is the adsorption oxygen, and the other is the lattice oxygen. These reactions of oxygen with CO are of the following types [29-32].

According to reaction of CO:

$$CO + 1/2 O_2 \rightarrow CO_2 \qquad (5)$$

for oxygen adsorption:
$$1/2 CO_2 + 2e^- \rightarrow O^{2-}_{ads} \quad (6)$$
$$O + e^- \rightarrow O^-_{ads} \quad (7)$$

for CO adsorption:
$$CO_g \rightarrow CO^+_{ads} + e^- \quad (8)$$
$$CO_g \rightarrow CO_{ads} \quad (9)$$

reaction of adsorbed species:
$$CO_{ads} + O^- \rightarrow CO_2 + e^- \quad (10)$$

forming a surface donor complex:
$$CO_g \rightarrow [OCOV_o]^+ + e^- \quad (11)$$

forming a surface vacancy
$$CO_g + O^x_o \rightarrow CO_2 + e^- + V^+_o \quad (12)$$

reaction with a surface complex:
$$1/2 O_2 + [OCOV_o]^+ + e^- \rightarrow CO_2 + O^x_o \quad (13)$$

reaction with a surface vacancy:
$$CO + 1/2 O_2 + V^x_o + e^- \rightarrow CO_2 + O^x_o \quad (14)$$

Reactions "(11)"- "(14)" include the lattice oxygen reaction and surface complex. In order to react with CO, the lattice oxygen vacancies must be continuously diffused to the surface, which occurs at higher temperature. For the vacancies to migrate from the bulk to the surface, Reactions "(11)"-"(14)" usually occur at temperature higher than 300°C [33].

The oxidation of CO is a process of thermal activation, but the desorption of oxygen is enhanced with the increase of temperature. So a comprised reaction rate of CO oxidation must take place at a proper temperature. In the present study, the optimized temperature was found to be 100°C. The films used for CO detection can be completely recovered by heating to a temperature of over 450°C.

7.2 Modified ZnO as H$_2$S Gas Sensor

The ZnO films where surface modified by dipping it into the solution of nano Ag. the nano Ag species would be distributed uniformly through out the surface of the film Due to this not only the initial resistance of the film is high but this amount is sufficient to promote the catalytic reaction effectively, and the overall change in the resistance on exposure of reducing gas (H$_2$S) leading to high sensitivity. When the ZnO film is exposed to H$_2$S gas, the reductive gas reacts with the oxygen adsorbed on the film surface. Then the electrons are released back in to the semiconductor, resulting in the change in the electrical conductance of ZnO film. It can be expressed in the following reaction

$$H_2S + 3O^{2-} \rightarrow H_2O + SO_2 + 6e^- \quad (15)$$

For the nano Ag/ZnO film, the low response at low operating temperature can be attributed to the low thermal energy of the gas molecules, which is not enough to react with the surface adsorbed oxygen species. As a result, the reaction rate between them is essentially low [27, 34] and low response is observed. On the other hand, the reduction in response after the optimum operating temperature may be due to the difficulty in exothermic gas adsorption at higher temperature [35]. There fore the maximum gas response can just be observed at the right operating temperature.

Figure12, illustrates the effect of dispersion of surface additive on the film conductivity. When amount of Ag on the surface of film is less than the optimum, the surface dispersion, may be poor and the sensitivity of the film is observed to be decreased. Since the amount may not be sufficient to promote the reaction more effectively. On the other hand, as the amount of Ag on the surface is more than optimum, the Ag atoms would be distributed more densely on the film. As a result, the initial resistance of the film would be decreases and the overall change in resistance on the exposure to gas would be smaller leading to lower response to the target gas.

Uniform and optimum dispersion of an additive dominates the depletion of electrons from semiconductor. Oxygen adsorbing on additive (misfits) removes electrons from the additive and additive in turn removes the electron from the near by surface region of the semiconductor and could control the conductivity.

For optimum dipping time (10min), the number of nano Ag misfits would be optimum and would disperse uniformly covering the complete film surface figure 12 (b). Adequate dispersion of nano Ag misfits (10min) on film surface would

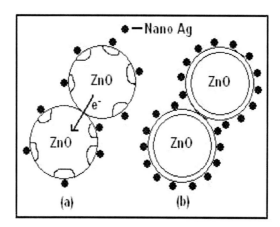

Fig. 12. Catalyst dispersion (a) poor and (b) adequate

produce depletion region on the grain surfaces and conductivity could be monitored systematically. The film conductivity would be very low in air and very high on exposure of H_2S gas and there fore, the gas response would be largest. For the dipping time smaller than the optimum, the number of nano Ag misfits would be smaller, their dispersion would be poor and the depletion regions would be discontinuous and there would be the paths to pass electrons from one grain to another figure 12(a). Due to this, the initial conductance (air) would be relatively larger and in turn, sensitivity would be smaller.

The introduction of nano Ag atoms on the surface of ZnO films by dipping technique increases the adsorption capability of the films. The nano Ag increases the surface to volume ratio of the film. The nano Ag particles on the surface control the grain boundary [37]. The different chemical identities on the surface of the film would alter the adsorption-desorption kinetics and gives unusual physical and chemical properties.

8 Summary

Following conclusions can be drawn from the experimental results:

a) The pure ZnO film showed highest response to CO gas (100ppm) at 100°C
b) The ZnO films were modified by nano Ag solution by dipping Technique. The surface modified films showed maximum response to H_2S gas (100ppm).
c) The surface modification shifts the response of film from CO to H_2S gas.
d) Due to introduction of nano Ag on the surface would alter the adsorption-desorption relationship if the film. The optimum dipping time was found to be 10min.
e) The surface modification alters only the surface morphology of the films not the bulk properties.
f) The surface modification facilitated adsorption of larger number of oxygen ions on the surface, could immediately oxidize the exposed H_2S gas leading to faster response time of the sensor.
g) The higher the oxygen deficiency, more the adsorption of oxygen ions on ZnO and hence faster would be the adsorption-desorption of target gas in result fast would be recovery of the sensor.
h) The modified ZnO thick film was observed to be highly selective to H_2S gas.
i) The quick response of the sensor could be attributed to larger oxygen deficiency in ZnO material.
j) A quick response and fast recovery were the special characteristics of the ZnO material.

Acknowledgments

The Author (MKD) is very much thankful to B.C.U.D., University of Pune, for financial assistance to this project, Sarchitnis, M.V.P. Samaj, Nashik and Principal, Arts, Commerce and Science College, Nandgaon, for his keen interest in this project The co-authors (VBG), (GHJ) are grateful to U.G.C., New Delhi, for financial assistance to this project and PKK thanks Department of Information Technology (DIT), New Delhi for financial support through grant no. 20(7) 2003VCND.

References

[1] Takao, Y., Miyazaki, K., Shimizu, Y., Egashira, M.: High ammonia sensitive semiconductor gas sensors with double layer structure and interface electrode. J. Electrochem. Sco. 141, 1028–1034 (1994)

[2] Nanto, H., Sokooshi, H., Kawai, T.: Aluminium –doped ZnO thin film gas sensors capable of detecting freshness of sea food. Sens. Actuators 14, 175–717 (1994)

[3] Seiyama, T., Kato, A., Fujiishi, K., Nagatani, M.: A new detector for gaseous components using semiconductive thin films. Anal. Chem. 34, 1502–1503 (1962)

[4] Yamazoe, N., Sakai, G., Shimaone, K.: Oxide semiconductor gas sensors. Catal. Surveys Asia 1, 63–75 (2003)

[5] Kwon, T., Park, S., Ryu, J., Choi, H.: Zinc Oxide thin film doped with Al_2O_3, TiO_2 and V_2O_5 as sensitive sensors for trimethylamine gas. Sen. Actuators 46, 75–79 (1998)

[6] Rao, G., Rao, D.: Gas sensitivity of ZnO based thick film sensors to NH_3 at room temperature. Sens. Actuators 55, 166–169 (1999)

[7] Basu, S., Datta, A.: Room temperature hydrogen sensors based on ZnO. Mater. Chem. Phy. 47, 93–97 (1997)

[8] Park, C.O., Akbar, S.A., Hwang, J.: Selective gas detection with catalytic filter. Mater. Chem. Phys. 75, 56–60 (2002)

[9] Sciliano, P.: Preparation, Characterisation and Application of thin films foe Gas sensors by cheap chemical method. Sens. Actuators 70, 153–164 (2000)

[10] Ogawa, H., Nishikawa, M., Abe, A.: Hall measurement and an electrical conduction model of Tin oxide ultra fine particle films. J. Appl. Phys. 53, 4448–4450 (1982)

[11] Yamazoe, N.: New approach for improving semiconductor gas sensors. Sens. Actuators 5, 7–19 (1991)

[12] de Lacy Costello, B.P.J., Ewen, R.J., Ratcliffe, N.M.: Highly sensitive mixed oxide sensors for the detection of ethanol. Sens. Actuators 87, 207–210 (2002)

[13] Penza, M., Martucci, C., Cassano, G.: No_x gas sensing characteristics of WO_3 thin film activated by noble metals (Pd, Pt, Au, Ag) layers. Sensors and Actuators 50, 52–59 (1998)

[14] Korotechnov, G., Brynzari, V., Dmitriev, S.: SnO_2 film for thin film gas sensor design. Mater. Sci. Eng. 56, 195–204 (1999)

[15] Servent, A.M., Rickerby, D.G., Horrillo, M.C., Saint-Jacques, R.G., Guitierrez, J.: Transmission electron microscopy investigation of SnO_2 thin films for sensor devices. Nanostruct. Mater. 11, 813–819 (1999)

[16] Kaciulis, S., Pandolfi, L., Viticoli, S., Sberveglieri, G., Zampiceni, E., Wlodarski, W., Galatsis, K., Li, Y.X.: Investigation of thin films of mixed oxides for gas-sensing applications. Surf. Interface Anal. 34, 672–676 (2002)

[17] Khanna, P.K., Singh, N., Kulkarni, D., Deshmukh, S., Charan, S., Adhyapak, P.V.: Water based simple synthesis of re-dispersible silver nano-particles. Materials Letters 61, 3366–3370 (2007)

[18] Roy Morrison, S.: Mechanism of semiconductor gas sensor operation. Sens. Actuators 11, 283–287 (1987)
[19] Khanna, P.K., Singh, N.D., Kulkarni, R., Marimuthu, S.: One-step preparation of nanosized Ag-Pd co-powder and its allformation at low temperature, Synthesis and Reactivity in Inorganic. Metal-Organic and Nano-Metal Chemistry 37, 1–9 (2007)
[20] Schierbaum, K.D., kirner, U.K., Geiger, J.F., Gopel, W.: Schottky- barrier and conductive gas sensors based upon Pd/SnO$_2$ and Pt/Tio$_2$. Sens. Actuators 4, 87–94 (1991)
[21] Jain, G.H., Gaikwad, V.B., Kajale, D.D., Chaudhari, R.M., Patil, R.L., Pawar, N.K., Deore, M.K., Shinde, S.D., Patil, L.A.: Gas sensing performance of pure and modified BST thick film resistor. Sensors and Transducers 90, 160–173 (2008)
[22] Jain, G.H., Gaikwad, V.B., Patil, L.A.: Studies on gas sensing performance of (Ba$_{0.8}$Sr$_{0.2}$) (Sn$_{0.8}$Ti$_{0.2}$) O$_3$ thick film resistors. Sens. Actuators 122, 605–612 (2007)
[23] Patil, D.R., Patil, L.A., Amalnerkar, D.P.: Ethanol gas sensing properties of Al$_2$O$_3$-dopped ZnO thick ilm resistors. Bull. Mater. Sci. 30(6), 553–559 (2007)
[24] Wagh, M.S., Jain, G.H., Patil, D.R., Patil, S.A., Patil, L.A.: Modified Zinc oxide thick film resistor as NH3 gas sensor. Sens. Actuaors B115, 128–133 (2006)
[25] Jain, G.H., Patil, L.A., Wagh, M.S., Patil, D.R., Patil, S.A., Amalnerkar, D.P.: Surface modified BaTiO$_3$ thick film resistors as H$_2$S gas sensors. Sens. Actuators 117, 159–165 (2006)
[26] Ishihara, T., Kometani, K., Hashada, M., Takita, Y.: Application of mixed oxide capacitor to the selective carbon dioxide sensor. J. Electrochem. Soc. 138, 173–175 (1991)
[27] Maosang, T., Dai, G.R., Gao, D.S.: Surfacemodification of oxide thin film and its gas sensing properties. Appl. Surf. Sci. 171, 226–230 (2001)
[28] Chu, X.F., Jiang, D.L., Guo, Y., Zheng, C.M.: Ethenol gas sensor based on CoFe2O4 nano crystallines prepared by hydrothermal method. Sens. Actuators 120, 177–181 (2006)
[29] Liu, Y.L., Wang, H., Liu, Z.M., Yang, H.F., Shen, G.L., Yu, R.Q.: Hydrogen Slfied sensing properties of NiFeO4 nanopowder doped with nobel metal. Sens. Actuators 102, 148–154 (2004)
[30] Göpel, W.: Chemisorptions and charge transfer at ionic semiconductor surface. Progr. Surf. Sci. 20, 9–103 (1985)
[31] Willett, M.J.: Spectroscopy of surface reactions. In: Mosely, P.T., Norris, J., Williams, D.E. (eds.) Techniques and Mechanism in Gas Sensing, Adam Hilger, Bristol (1991)
[32] Lampe, U., Garblinger, M.J.H.: Carbon monoxide sensor based on thin films of BaSnO$_3$. Sens. Actuators 25, 657–660 (1995)
[33] Kabayashi, T.M., Haruta, S.H., Nakane, M.: A selective CO sensing Ti-doped α-Fe$_2$O$_3$ with co-precipitated ultrafine particles of gold. Sens. Actuators 13, 339–348 (1988)
[34] Hu, Y., Tan, O.K., Cao, W., Zhu, W.: Fabrication and characterization of nano-sized SrTiO$_3$-based oxygen sensor for near room-temperature operation. Sensors 5, 825–832 (2005)
[35] Chang, J.F., Kuo, H.H., Leu, I.C., Hon, M.H.: The effect of thickness and operation temperature on ZnO –Al thin film CO gas sensor. Sens. Actuators 84, 258–264 (2002)
[36] Yamazoe, N.: New approaches for improving semiconductor gas sensors. Sens. Actuators 5, 7–19 (1991)

Micro Temperature Sensors and Their Applications to MEMS Thermal Sensors

Mitusteru Kimura

Tohoku gakuin university,
Tagajo, Miyagi, 985-8537, Japan

Abstract. New type thermistor-like absolute temperature sensors as a diode-thermistor and a transistor-thermistor, and a temperature sensor proportional to the absolute temperature, which are proposed by the author, are presented. Novel temperature-difference sensor based on the short circuit Seebeck-current measurement of a single thermocouple using the imaginary short circuit of the OP amplifier is also proposed, and a method to evaluate the performance of the thermal sensor by use of the generated power due to the IR absorption is presented. This temperature difference sensor is usually used combined with a absolute temperature sensor. This new temperature difference sensor is first realized as an IR sensor combined with the diode-thermistor. This short circuit Seebeck-current measurement type single thermocouple is also applied to the vacuum sensor. We have extended measurable pressure range of the thin film Pirani vacuum sensor that is still sensitive above 1 atm. In our thin film Pirani vacuum sensor, the proposed temperature difference sensor is used in order to get extremely high sensitivity, especially in high vacuum range. We have achieved much wider measurable pressure range over 8 digits by use of our new thin film Pirani vacuum sensor than that of the traditional one. We could expand the pressure sensitivity beyond 1 atm. by adoption of the vibration of the sensing cantilever due to the sudden heating.

1 Introduction

It is very important to reduce both the thermal capacity and the thermal conductivity to improve the time response and the sensitivity for thermal sensors [1]. Therefore, MEMS with thermal sensors adopts a thermally isolated hot plate and temperature sensors from the substrate.

The author proposed a micro air–bridge heater fabricated by using silicon planar technology and micromachining of silicon, and suggested that this micro air–bridge heater with a temperature sensor will be used as a flow sensor, thermal vacuum sensor, etc. to improve their sensitivity due to larger interaction area with the air and thermal isolation of the hot suspended film [2], [3]. Then the author has proposed absolute temperature sensors [4]-[7] and a temperature difference sensor [8] for MEMS sensors. Generally, the temperature difference sensor is used combined with an absolute temperature sensor to monitor the base temperature.

In this paper above mentioned newly developed temperature sensors and their applications to MEMS thermal sensors are presented.

2 Principle of Newly Developed Temperature Sensors

2.1 Transistor-Thermistor and Diode-Thermistor

In bipolar transistor it is deduced that the collector current Ic has the same temperature T dependence as the NTC thermistor under the constant emitter-base voltage Ve for qVe, qVc>> nkT, as shown below [4],

$$I_c = I_\infty \exp\left(-\frac{q(V_d - V_e)}{nkT}\right) + I_{ceo}, \qquad (1)$$

where I_∞ is the collector current at T=∞,
q is electronic charge,
V_d is the diffusion potential,
n is the ideal factor,
k is Boltzmann constant,
I_{ceo} is the collector current at base terminal opened.

Generally, I_{ceo} is relatively small, we can omit this factor. Therefore, we can rewrite equation (1) as follows:

$$\log I_c = -B\frac{1}{T} + C \qquad (2)$$

$$B = q\frac{(V_d - V_e)}{nk}. \qquad (3)$$

Diffusion potential V_d of the pn junction between the emitter and the base has a weak dependence on the temperature if the junction is composed of materials with relatively high impurity concentration. Equation (2) shows the same temperature T dependence as the NTC thermistor, and B is equivalent with the thermistor constant, in addition, we can see from equation (3) that this B changes with a change of Ve.

In figure 1 the schematic diagram of the npn bipolar transistor with suitable applied voltages working as the thermistor-like transistor is shown.

Since the collector current Ic is almost the same as the emitter current Ie and thermistor-like temperature effect is determined by the barrier height, $q(V_d-V_e)$, of the pn junction between the base and the emitter as shown in figure 1 and equation (3), we notice that the pn junction diode will also have the same temperature effect as the thermistor-like transistor [4]. Therefore, the author has named the thermistor-like transistor the transistor-thermistor and the thermistor-like diode the diode-thermistor.

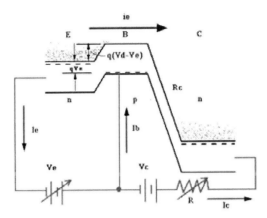

Fig. 1. Schematic diagram of the npn bipolar transistor with suitable applied voltages working as a thermistor-like transistor.

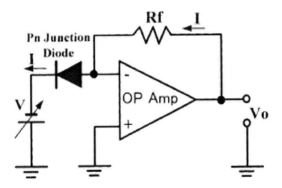

Fig. 2. Experimental circuits for the diode to work as a diode-thermistor.

It is important to fix the forward bias applied voltage to the pn junction, such as Ve for the transistor-thermistor, in order to maintain the thermistor-constant B shown in equation (3). The transistor-thermistor has a merit that the bias voltage application to set the thermistor-constant B to be a suitable value is easy, because the transistor has three terminals and, therefore, the suitable Ve is able to set independently on measuring Ic, which produces the output voltage across the load resistance. On the other hand, it is hard to set the fixed applied voltage to the diode-thermistor, because the diode has only two terminals. Therefore, we have adopted the bias application circuits as shown in figure 2 to overcome the problem. However, the diode-thermistor has a merit to be compact and easy to form on a MEMS structure, such as a cantilever.

The temperature sensitivity, corresponding to B in equation (3), of this pn junction diode-thermistor can be adjusted by the forward bias-voltage V, since the diode current I of this sensor has an exponential factor with respect to the reciprocal absolute temperature T at a constant bias-voltage V.

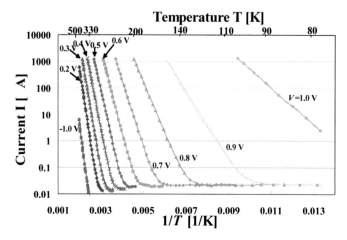

Fig. 3. Characteristics of the diode-thermistor, in which a bipolar transistor works as a pn diode by means of shortening their base and collector.

These transistor-thermistor and diode-thermistor have more sensitive and different from the usual diode temperature sensor, such as the thermodiode, which is based on the temperature dependence of the forward voltage change at the constant forward current, while our proposed transistor-thermistor and diode-thermistor are based on the temperature dependence of the forward current change at the constant or fixed forward voltage applied to the pn junction.

Characteristics of this diode-thermistor, in which a bipolar transistor works as a pn diode by means of shortening their base and collector, are shown in figure 3. We can see that we can expand the measurable temperature range, such as 77K-500K for Si diode, as a thermistor-like absolute temperature sensor by choose the suitable forward and fixed applied voltage.

2.2 Diode Temperature Sensor Proportional to the Absolute Temperature

Traditional thermistor and thermdiode are very familiar as an absolute temperature sensor. However, the thermistor has a narrow measurable temperature range as a single temperature sensor and has a non-linear relationship between the temperature T and the output voltage. As to the thermdiode [9],[10] and the IC temperature sensor[11],[12], they have a linear dependence between forward voltage V and absolute temperature T under a fixed forward currrent I_0, however, it needs the off-set balance control in order to amplify the dc signal. The principle of the thermodiode is based on equation (1) for the diode current I and the forward voltage V instead of Ic and Ve in the transistor, respectively.

From this equation (1) we can get the following equation,

$$V = V_d - \frac{nkT}{q}\ln\frac{I_\infty}{I}. \qquad (4)$$

In figure 4 the relationship between the forward voltage V of a pn junction diode and the absolute temperature T predicted from equation (4) is shown. This is the principle of the traditional thermodiode.

Micro Temperature Sensors and Their Applications to MEMS Thermal Sensors

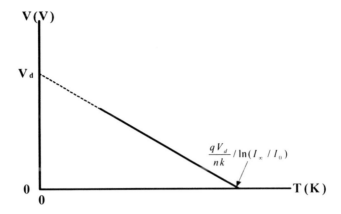

Fig. 4. Relationship between the forward voltage V and the absolute temperature T of the traditional thermodiode.

On the other hand, in our proposed temperature sensor, we can get the following equation from equation (4),

$$\Delta V = V_2 - V_1 = \frac{nkT}{q}(\ln\frac{I_\infty}{I_1} - \ln\frac{I_\infty}{I_2}) = \frac{nkT}{q}\ln(1+\frac{\delta I}{I_1}) = \beta T, \quad (5)$$

$$\beta = \frac{nk}{q}\ln(1+\frac{\delta I}{I_1}), \quad (6)$$

where $\delta I = I_2 - I_1$.

For both $I_1 = I_0$ and δI being constant, β in equation (6) will be also constant.

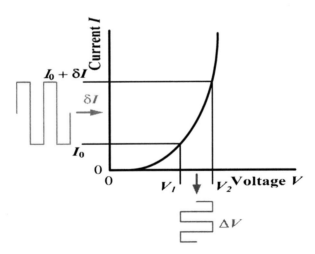

Fig. 5. Schematic diagram of a superimposed ac signal of a constant δI to the dc forward constant current I_0 and the ac output signal voltage ΔV for a pn junction diode.

If we use superimposed ac signal of a constant δI to the dc forward constant current I_0 and pickup the ac output signal voltage ΔV via the dc-cut capacitor connected to the pn junction diode as shown in figure 5 and 6, we can get the output voltage ΔV proportional to the absolute temperature T as shown in figure 7, in where the output voltage ΔV is amplified by an OP amplifier and its output voltage V_0 is shown as a vertical axis. It has merits that the calibration of our temperature sensor should be done at only one temperature point, that temperature measurement by use of ac signal can be isolated from dc circuits for bias voltage application circuits, and that thermometry of a very small area can be achieved because of a simple and very small sensing probe of a pn junction diode [13].

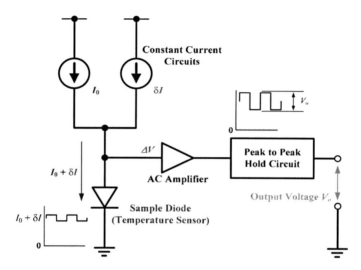

Fig. 6. Schematic diagram on the principle of the diode temperature sensor proportional to the absolute temperature.

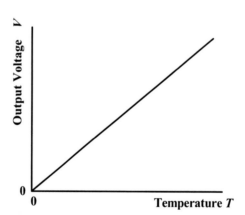

Fig. 7. Schematic diagram of expected characteristics of the diode temperature sensor proportional to the absolute temperature.

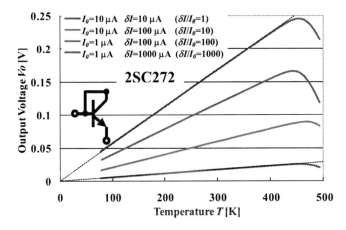

Fig. 8. Experimental results of the output voltage ΔV (shown as an amplified voltage V₀) proportional to the absolute temperature T for a pn junction diode (base and collector of a transistor are shortened).

In figure 8 the experimental results using a pn junction diode between the emitter and the base of the transistor, 2SC2720, in which the base and the collector are shortened, are shown. We can see that very good relationships as a temperature sensor proportional to the absolute temperature T are obtained. It can measure the wide temperature range such as 77-400K. We have ascertained that the Schottky barrier diode has also similar characteristics, although it has less measurable maximum temperature than the pn junction diode.

2.3 Principle of the Short Circuit Seebeck-Current Measurement Type Thermocouple

It is well known that the Seebeck coefficient α_s for silicon is approximated as a function of electrical resistivity ρ as shown below [14]:

$$\alpha_s = \frac{mk}{q} \ln \frac{\rho}{\rho_0}, \tag{7}$$

where $\rho_0 \approx 5 \times 10^{-4}$ Ωcm,

m = 2.6 for Si,

k is Boltzmann constant.

We have noticed that it is easy to decrease the resistivity three or four orders, which can extremely increase the short circuit Seebeck-current of the thermocouple, while the Seebeck coefficient α_s decreases only several times [15]. If we adopt the method of the short circuit Seebeck-current measurement using a single thermocouple with the n[++]-Si and metal film in stead of the method of the open-circuit voltage measurement using the traditional thermopile, it is expected

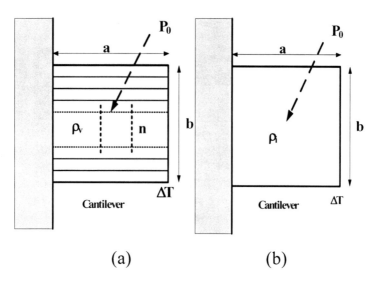

Fig. 9. Structural models of thermal IR sensors with the same IR sensing area of cantilever. (a) Traditional n-pairs thermopile, (b) Proposed single thermocouple for the Seebeck-current measurement type.

that our proposed single thermocouple with extremely low resistance can generate higher thermoelectric power W_i than W_v of a traditional open-circuit thermopile, where subscript i and v correspond to current measurement (single thermocouple) and voltage measurement (traditional thermopile), respectively [16].

In figure 9, theoretical models of the micro-air-bridge (MAB) type thermal sensors are shown; one is a thermopile model composed of n-pairs of thermocouples (a) and the other is a single thermocouple (b). Both (a) and (b) have the same area (a × b) of IR absorption region, and is assigned the same temperature gradient ΔT due to the absorption of incident IR power P_0. And they have different resistivity ρ_v and ρ_i. Internal resistance r_v and r_i of sensors are proportional to their resistivity ρ_v and ρ_i, respectively.

In the thermopile (a), following equations are obtained: the total thermoelectric voltage $E_v = n\, e_v = n\alpha_v\, \Delta T$; internal resistance $r_v = n\, r_{0v}$ (r_{0v}: resistance of each thermocouple); generated power $W_v = E_v^2 / r_v$. In the single thermocouple (b) following equations are also obtained: for $\gamma = \alpha_v /\alpha_i$ and $\rho_i /\rho_v = 10^{-f}$, the thermoelectric voltage $E_i = e_i = \alpha_i\, \Delta T$; internal resistance $r_i = r_{0v} \times 10^{-f} / n$; generated power $W_i = E_i^2 / r_i = \alpha_i^2 \Delta T^2 / r_i$.

Therefore, it is derived the generated thermoelectric power ratio M of W_i and W_v by following equation:

$$M = W_i / W_v = 10^f / \gamma^2 = \rho_v\, \alpha_i^2 / (\rho_i\, \alpha_v^2). \tag{8}$$

From equation (8), we can see that the ratio of the thermoelectric power by each sensor is only the function of the Seebeck coefficient and the resistivity. We can notice that equation (8) has the same meaning as the ratio of the Figure of Merit,

$Z=\alpha_s^2/\rho\kappa$, in the thermoelectric materials, where κ is the thermal conductivity, because κ is independent of the impurity concentration in the Si material, and is cancelled in equation (8).

For f =3, M will be 92 because of γ=3.3. For example, it is seen that the generated thermoelectric power, W_i, of a single thermocouple with extremely low resistivity semiconductor such as n^{++}-Si thin film (e.g. $\rho_i = 10^{-2}$ Ωcm) is expected to be 92 times higher than W_v of the thermopile with n-pairs thermocouples made of relatively high resistivity Si (e.g. ρ_v = 10 Ωcm), because of f = 3 according to the definition on the ratio of ρ_v and ρ_i. And we can see that M will be one for f = 0 ($\rho_v = \rho_i$) in the case of γ=1. It means that both n-pairs thermopile and our proposed single thermocouple have the same generated thermoelectric powers as long as they are made of same materials with the same resistivity. Namely, they are equivalent. However, since each pair of thermocouple of the thermopile requires line and space, the internal resistance will be larger than an ideal n-pairs thermopile configuration shown in figure 9 (a).

We can use the imaginary short of the OP amplifier to measure the short circuit Seebeck-current of the single thermocouple sensor as shown in figure 10. The combination of the single thermocouple sensor and the OP amplifier will compose temperature difference sensor.

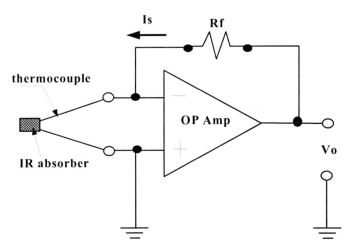

Fig. 10. Measuring circuit for short circuit Seebeck-current of a single thermocouple using imaginary short of an OP amplifier.

The output voltage Vo of the OP amplifier shown in figure 10 is given as,

$$V_o = R_f I_s = R_f V_t / r, \qquad (9)$$

where Is is the short circuit Seebeck-current,

Rf is the feedback resistance of the OP amplifier circuit,

r is the internal resistance of the thin film thermocouple.

So we can get the temperature difference ΔT as follows,

$$\Delta T = V_o r / \alpha_s R_f . \tag{10}$$

If we measure the internal resistance of the thin film thermocouple beforehand, we can see the temperature difference ΔT from the above equation (10). We have measured the Seebeck coefficient α_s of the phosphorus diffused Si layer (n^{++}-Si) at 1100°C for 2 hrs, and found that it is about -160 μV/K.

3 Applications of the Short Circuit Seebeck- Current Measurement Type Thermocouple

3.1 Micro-Air-Bridge (MAB) Type IR Sensor Composed of a Single Thermocouple

The IR sensor with a micro-air-bridge (MAB) type sensing area composed of a single thermocouple of a double layer of n^{++}-Si and Ni metal (a very thin film Cr is deposited as an adhesion layer between Ni and SiO_2 film) is proposed [16]. Four beams, which consist of the double layer of the n^{++}-Si and Ni metal of a thin film as a thermocouple, are electrically parallel-connected each other to reduce the electrical internal resistance r_s of the thermocouple. In this IR sensor Ni is adopted as a pair conductor material for the n^{++}-Si of the thermocouple because of opposite sign of Seebeck coefficient. SOI (6.5 μm thick) substrate is used as a starting material, and the heavily doped n^{++}-Si layer is formed. The thickness of n^{++}-Si diffused layer (assuming an average impurity concentration of about 1×10^{20} cm^{-3} in this layer) is estimated to be about 2 μm thick. In this case the hot junction of the thermocouple is located at the centre of the MAB being the main IR sensing area and the cold junction is located at the SOI substrate as shown in figure 11. Dimension of the MAB sensing area is 700 μm x 700 μm and about 6.5μm thick. Each beam is about 380 μm long and about 200 μm wide.

The gold-black layer as an IR absorber is deposited on the IR sensing region including the four beams through the slit of the metal mask in the final process step. In figure 12 a micrograph of the IR sensor chip with the gold black as a IR absorber and the diode-thermistor is shown. The diode-thermistor is formed on the Si substrate to be able to monitor the base temperature of the thermocouple.

In figure 13 the output signal of our proposed IR sensor of the MAB type of a single thermocouple is shown for the received IR power of 40.6 μW. In this case the IR received area (sensing area) is supposed to be about 7.9×10^{-3} cm^2 because of total area of the MAB including the four beams, and the set-up of measurement of IR detection. We can see that the output voltage of the OP amplifier through the feedback resistance of 100 kΩ is 56mV. The internal resistance of this MAB type IR sensor was about 20 Ω, therefore, the resistance of each beam will be about 80 Ω because of four electrically parallel-connected beams.

We will compare the performance of this MAB type IR sensor in terms of the generated thermoelectric power with that of the commercial one.

Micro Temperature Sensors and Their Applications to MEMS Thermal Sensors 319

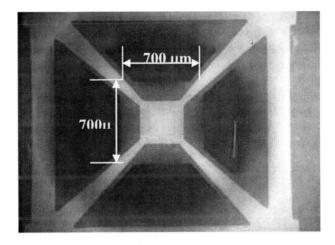

Fig. 11. MAB type IR sensor using the method of the short circuit Seebeck-current measurement of a single thermocouple.

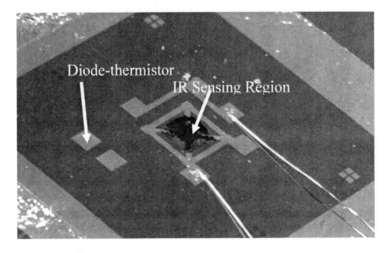

Fig. 12. Micrograph of the IR sensor chip with the gold black as a IR absorber and the diode-thermistor.

The commercial thermopile composed of n^{++}-poly-silicon /Al film (made by a Japanese company) has the internal resistance r_v of 65 KΩ, responsivity R_s of 15 V/W and thermal time constant τ of 15 msec. We have notice that the commercial thermopile is formed using almost the same materials as our proposed short circuit Seebeck-current measurement type IR sensor. The commercial thermopile do not adopt the p type semiconductor to reduce its internal resistance because the heavily doped p type Si have one order larger resistivity than that of the heavily doped n type one.

The output voltage of the OP amplifier for the received IR power P_i of 40.6 µW is about 56 mV as shown in figure 13. The short-circuit Seebeck-current of the MAB IR sensor due to P_i = 40.6 µW is estimated to be about I_i = 0.56 µA taking account of the R_f = 100kΩ in figure 10. We can calculate the generated power W_i = $r_s I_i^2$ = 6.27×10^{-12} W due to P_i taking account of internal resistance r_i =20 Ω. It is important to notice W_i being independent on the OP amplification. This internal resistance is larger than that as expected value and this mean that the average thickness of impurity diffused n^{++}-Si layer into the SOI layer is relatively thin and results in less than 2µm.

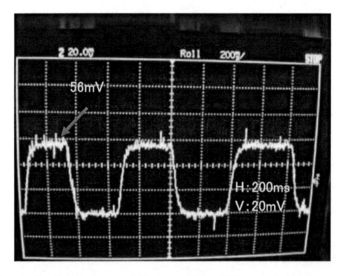

Fig. 13. Output signal of the proposed IR sensor of the MAB type of a single thermocouple.

The commercial thermopile will have the output voltage of E_v = Rs·Pi = 6.09×10^{-4} V for the same P_i of 40.6 µW. Therefore, we can see the generated thermoelectric power output W_v (= E_v^2 / r_v) by this commercial thermopile itself is 5.7×10^{-12} W taking account of its internal resistance r_v = 65 KΩ. After all, we can obtain the ratio of generated thermoelectric powers, M= W_i / W_v ≈1.1. It means that the performance of our prototype single thermocouple is equivalent to the commercial excellent thermopile. However, our proposed sensor has great advantages such as dramatically simple fabrication process resulting in low cost. Moreover it is expected that the performance of our prototype thermocouple can be further improved by adoptions of thinner SOI cantilever with all n^{++}-Si layer to reduce the internal resistance r_s and thermal loss, and better IR absorption layer.

3.2 New Type Thin Film Vacuum Sensor

We have planned to adopt the temperature difference sensor in order to get the higher sensitivity in the higher vacuum range. Our proposed short circuit

Seebeck-current measurement type thermocouple has more merits than the traditional thermopile due to easy fabrication, compactness and sensitivity.

In figure 14 the schematic diagram of the new type thin film vacuum sensor composed of the microheater and two temperature difference sensors of the short circuit Seebeck-current measurement type thermocouple formed on the SOI cantilever is shown. The microheater is made of the sputtering deposited Nichrome film and the thin film thermocouple is composed of the phosphorus diffused Si layer (n^{++}-Si) and a Ni/ Nichrome double layer.

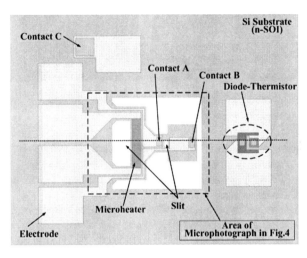

Fig. 14. Schematic diagram of the new type thin film vacuum sensor composed of the microheater and two temperature difference sensors of the short circuit Seebeck-current measurement type thermocouple.

In figure 15 the micrograph of the sensing region, which consists of a cantilever with two thin film thermocouples, of the new type thin film vacuum sensor is shown.

We have used the n-SOI substrate ((100), $7.8/1.0/550\mu$m, 3.00-$5.00/40.0$ Ωcm) as a starting material. After initial cleaning, the cantilever pattern is formed by chemical etching of the SOI layer and the thermal silicon dioxide film (SiO_2 film) is formed by thermal oxidation. After the lithography of thermal silicon dioxide film (about 0.5μm thick) to open windows for the low resistivity (heavily doped) n^{++}-Si regions ($<10^{-3}$ Ωcm), thermal diffusion using phosphorous OCD-coating diffusion source at $1100°$C for 1 hour is carried out. Then the thermal diffusion for p^{+} region formation of pn diode is carried out using the boron diffusion source of the OCD at $1100°$C for 1 hour. The thickness of n^{++}Si diffused layer (assuming an average impurity concentration of about 1×10^{20} cm^{-3} in this layer) is estimated to be about 1 μm thick. The Ti/Nichrome/Ni metal (total thickness: about 0.3 μm) of a pair material in a single thermocouple is deposited by RF sputtering and patterned using the etching process. The thin film thermocouple is composed of

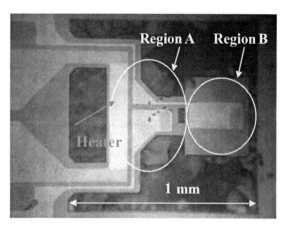

Fig. 15. Micrograph of the sensing region, which consists of a cantilever with two thin film thermocouples, of the new type thin film vacuum sensor.

the double layer of n^{++}-Si and Ti/Nichrome/Ni metal film electrically isolated via the thermal SiO$_2$ film (about 0.5 μm thick) formed on the SOI layer. The backside cavity to release the cantilever from the SOI substrate is formed using the anisotropic etchant of hydrazine for Si.

The microheater is formed on the SiO$_2$ layer on the SOI layer in order to isolate from the thermocouple, and the top deposited layer of the Ni film is etched away to increase the electrical resistance resulting in about 30 □. Slits are formed in the SOI cantilever in order to make thermal resistance both between the substrate and microheater, and between the region A and the region B.

In figure 16 the measuring circuits for the new type thin film vacuum sensor using the short Seebeck-current detection type thermocouple is shown.

Two thermocouples (thermocouple A and thermocouple B) of a pair of the n^{++}-Si layer and Ti/Nichrome/Ni metal film (mainly Ni thin film due to the low electrical resistance) are formed on the SOI cantilever. The thermocouple A can measure the temperature difference between the point in the region A, on where the microheater is formed, and the Si substrate, and the thermocouple B can do between the region A and the point in the region B thermally isolated from the region A by the slit. The cold junction and the hot junction of the thermocouple B will be the junction formed in the region B and the junction (region) formed in the region A, respectively. The same Ni (Ti/Nichrome/Ni) metal film lead-line is extended from the Ohmic junction point with the n^{++}-Si layer in the region A to the Si substrate rim to form the hot junction for the thermocouple B.

Temperature difference between the region A and the thermally isolated region B will be zero under the higher vacuum due to the cantilever structure. Therefore, the Seebeck current of the thermocouple B will be zero, namely the output voltage of the OP amplifier from the thermocouple B shown in figure 15 will also be zero under the high vacuum such as 1×10^{-4} Pa.

Micro Temperature Sensors and Their Applications to MEMS Thermal Sensors 323

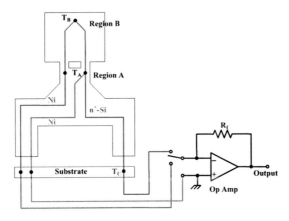

Fig. 16. Measuring circuits for the new type thin film vacuum sensor using the short Seebeck-current detection type thermocouple.

The microheater is driven by constant current of about 40 mA, and the increased temperature was about 190°C at this constant current under the high vacuum.

In our vacuum sensor, the output voltage has remained by only about 10 mV for $R_f = 100$ kΩ even under 1×10^{-4} Pa. Therefore, we have subtracted the remaining output voltage value from the data obtained from the vacuum pressure measurement.

Figure 17 shows output waveforms obtained from the thermocouple B when microheater is driven as heating for 100 msec and cooling for 100 msec. The vacuum pressure is measured at the measurement point P1 for lower pressure than about 0.5 atm. using the measurement circuits shown in figure 17.

In figure 18 the logarithmical scaled characteristics of the output voltage V_o vs. the vacuum pressure p, measured at P1 in figure 17, is shown. We can see that we can measure the wide pressure range of about $2 \times 10^{-3} - 1 \times 10^5$ Pa (1 atm.), which can

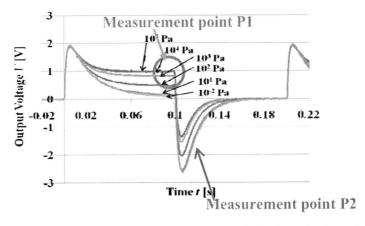

Fig. 17. Output waveforms obtained from the thermocouple B when microheater is driven as heating for 100 msec and its cooling for 100 msec.

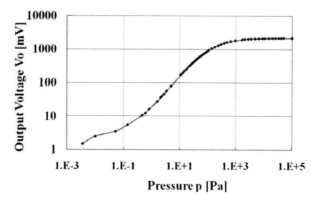

Fig. 18. Logarithmical scaled characteristics of the output voltage V_o vs. the vacuum pressure p, measured at P1 shown in figure 17.

be seen in the characteristics with linear plots (not shown here), by use of our vacuum sensor with our newly developed short circuit Seebeck-current measurement type thermocouples.

In figure 19 the schematic side view of the thin film Pirani vacuum sensor developed for high pressure measurement, in which the sensing cantilever is vibrating due to heating and cooling of the microheater, is shown. This measurement method is based on the air convection cooling due to the vibration of the thin film cantilever composed of double layers of a SOI layer and a BOX (SiO_2) layer, on which a microheater and two temperature-difference sensors are formed. The cantilever will vibrate when it is heated up and cooled down based on its "on" and "off" states, like a bimetal thermal bending, because double layers of the cantilever have very large difference in their thermal expansion coefficients. In our experiments the microheater is driven so as to be heating for 15 msec and cooling also for 20 msec in order to repeatedly heat up to about 190°C of the saturated temperature at 100mW under the higher pressure than near 1 atm. In this experiment with sensing-cantilever vibration the maximum amplitude of the output voltage Vo at P2 shown in figure 17, which is corresponding to the maximum temperature difference between the A and the B thermocouple, are measured after the microheater is turned off, because we can acquire the stable and reproducible pressure data due to vibration effects of the sensing cantilever.

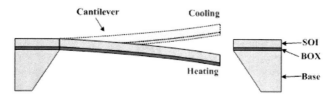

Fig. 19. Schematic side view of the thin film Pirani vacuum sensor with vibration-driving for high pressure measurement.

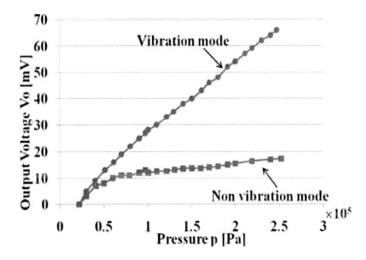

Fig. 20. Output characteristics of the thin film Pirani vacuum sensor with cyclic vibration mode and without vibration mode.

In figure 20 the output characteristics of the thin film Pirani vacuum sensor with the cyclic vibration mode and the non vibration mode are shown. We can see that the output voltage in high pressure range near 1 atm shows saturation characteristics as shown in figure 18, however, as shown in figure 20 increase in the output voltage V_o of the vibration mode is observed as increasing the pressure p, while the decrease in that of the non vibration mode is observed up to the pressure of 2.6 atm. These effects with the vibration mode may be due to heat transfer by the forced convection through the surrounding high pressure air.

4 Conclusion

New type thermistor-like absolute temperature sensors, diode-thermistor and transistor-thermistor, and a temperature sensor proportional to the absolute temperature, proposed by the author, are presented.

Novel temperature-difference measurement method based on short circuit Seebeck-current measurement of a single thermocouple with extremely low resistivity is also proposed, and it is demonstrated that the single thermocouple is superior or equivalent to the traditional thermopile through the theoretical model and the experiments. The authors believe that our proposed current measurement thermocouple sensor combined with the OP amplifier (as an active sensor) will be a promised one to be used mostly instead of a traditional thermopile in thermal measurement systems, such as IR sensor, uncooled IR imager, flow sensor and surface temperature-distribution sensors array, etc. because of its simple structure, high sensitivity and very small size. The single thermocouple sensor as a temperature-difference sensor will be more useful when it is combined with the diode-thermistor, fabricated on the same Si chip, as an absolute temperature sensor, because all of these sensors and circuits are full compatible to CMOS technologies.

We have proposed the thin film vacuum sensor that has a cantilever structure with the temperature difference sensors of the short circuit Seebeck-current measurement type thermocouple. In our experiments using the prototype sensor, very wide vacuum pressure range between 10^{-3} Pa and 10^5 Pa can be reproducibly measured. We could expand the pressure sensitivity range from 1×10^5 Pa to 2.6×10^5 Pa by adoption of the forced vibration of the sensing cantilever.

References

[1] Dereniak, L.E., Crowe, D.G.: Optical Radiation Detectors, ch. 6, p. 133. Wiley, NewYork (1984)
[2] Kimura, M.: Heater, JP. Application No. 1979-027559 (JP No. 1398241)
[3] Kimura, M.: Microheater and Microbolometer using Microbridge of SiO2 Film on Silicon. Electron. Letters 17, 80–82 (1981)
[4] Kimura, M.: Thermal Microsensors. Sensors 6, 257–275 (2000)
[5] Kimura, M., Toshima, K.: Thermistor-like pn junction temperature-sensor with variable sensitivity and its combination with a micro-air-bridge heater. Sensors and Actuators A 108, 239–243 (2003)
[6] Kimura, M.: Method and Apparatus for Temperature Measurement and Thermal Infrared Image Sensor (Transistor-Thermistor), JP publication No. 1999-287713 (JP No. 3366590)
[7] Kimura, M.: Method and Apparatus for Temperature Measurement and Thermal Infrared Image Sensor (Diode-Thermistor), JP publication No. 2001-264176 (JP No. 3583704)
[8] Lee, S.S., Kimura, M.: Short-Circuit Measurement by Seebeck Current Detection of A Single Thermocouple and its Application. Sensors and Actuators A 139, 104–110 (2007)
[9] Stemme, G.: A CMOS integrated silicon gas-flow sensor with pulse-modulated output. Sensors and Actuators 14, 293–303 (1988)
[10] Jin-Biao, H., Qin-Yi, T.: Integrated multi-function sensor for flow velocity temperature and vacuum measurements. Sensors and Actuators 19, 3–11 (1989)
[11] Gardner, J.W.: Microsensors, ch. 5, p. 100. Wiley, New York (1994)
[12] Bianchi, R.A., Vinci Dos Santos, F., Karam, J.M., Coutois, B., Pressecq, F., Sifflet, F.: CMOS compatible temperature sensor based on the lateral bipolar transistor for very wide temperature range applications. In: Proceedings of the 3rd THERMINIC Workshop, pp. 205–210 (1997)
[13] Takashima, N., Kimura, M.: Investigation on the Diode Temperature-Sensorwith the Output Voltage Proportional to the Absolute Temperature. IEE J. Trans. SM. 127, 328–332 (2008)
[14] Van Herwaarden, A.W., Sarro, P.M.: Thermal sensors based on the Seebeck effect. Sensors and Actuators A 10, 321–346 (1986)
[15] Kimura, M.: Detection method of temperature difference, temperature sensor and infrared sensor using the same, JP. Application No. 2004-026247 (JP publication No. 2005-221238)
[16] Kimura, M., Lee, S.S.: Proposal of a Temperature-Difference Sensor based on Short-circuit Current Measurement of a Single Thermocouple. In: Proc. ThETA1 (IEEE:07EX1655), January 2007, pp. 267–272 (2007)

Author Index

Abbott, Derek 169, 241
Abe, M. 15
Akeila, Ehad 203
Al-Sarawi, Said F. 169, 241
Alippi, C. 143

Bailey, Donald G. 91
Baste, Y.R. 283
Biwa, S. 15
Borhade, A.V. 283
Bouganis, Christos-Savvas 91
Brooker, Graham 271

Camplani, C. 143
Carnegie, D.A. 113
Chavan, D.N. 293
Chen, Liang-Yih 257
Chuang, Cheng-Hsin 189
Cree, M.J. 113

Deore, M.K. 293
Dissanayake, Don W. 241
Dorrington, A.A. 113

Enaka, Toshihiro 131

Gaikwad, V.B. 283, 293
Galperti, C. 143
Gomez, Jairo Alejandro 271
Gooneratne, C.P. 1
Guelaz, R. 55

Hashizume, Hiromichi 35

Ihara, I. 75
Ito, Toshio 35
Iwahara, M. 1

Jain, G.H. 283, 293
Jongenelen, A.P.P. 113

Kajale, D.D. 283, 293
Kakikawa, M. 1
Khanna, P.K. 293
Kimura, Mitusteru 309
Koschmieder, F. 221
Kourtiche, D. 55
Kurnicki, A. 1

Matsumoto, E. 15
Mukhopadhyay, S.C. 1

Nadi, M. 55

Payne, A.D. 113

Röck, H. 221
Rouane, A. 55
Roveri, M. 143

Salcic, Zoran 203
Sato, Hirofuji 131
Sato, Tetsuya 35
Shinde, S.D. 293
Sugimoto, Masanori 35
Swain, Akshya 203

Takahashi, M. 75
Takeno, Junichi 131
Tikka, Ajay C. 169
Tulathimutte, Kan 35

Wang, Yun 257

Yamada, H. 75
Yamada, S. 1
Yeow, John T.W. 257

Breinigsville, PA USA
19 November 2009
227839BV00010B/89/P